Springer Series in Molecular Biology

Series Editor: Alexander Rich

Springer Series in Molecular Biology

Series Editor: Alexander Rich

Roger L.P. Adams
Roy H. Burdon

Molecular Biology of DNA Methylation

With 88 Figures

Springer-Verlag
New York Berlin Heidelberg Tokyo

Roger L.P. Adams
Department of Biochemistry
University of Glasgow
Glasgow G12 8QQ
United Kingdom

Roy H. Burdon
Department of Bioscience and Biotechnology
University of Strathclyde
Glasgow G4 ONR
United Kingdom

Series Editor:
Alexander Rich
Department of Biology
Massachusetts Institute of Technology
Cambridge, Massachusetts 02139 U.S.A.

Library of Congress Cataloging in Publication Data
Adams, R.L.P. (Roger Lionel Poulter)
  Molecular biology of DNA methylation.
  (Springer series in molecular biology)
  Bibliography: p.
  Includes index.
  1. Deoxyribonucleic acid—Metabolism.
2. Methylation. I. Burdon, R.H. (Roy Hunter)
II. Title. III. Series.
QP624.A33   1985   574.87'3282   85-9883

Typeset by Bi-Comp Incorporated, York, Pennsylvania.

9 8 7 6 5 4 3 2 1

ISBN-13: 978-1-4612-9576-1   e-ISBN-13: 978-1-4612-5130-9
DOI: 10.1007/978-1-4612-5130-9

Series Preface

During the past few decades we have witnessed an era of remarkable growth in the field of molecular biology. In 1950 very little was known of the chemical constitution of biological systems, the manner in which information was transmitted from one organism to another, or the extent to which the chemical basis of life is unified. The picture today is dramatically different. We have an almost bewildering variety of information detailing many different aspects of life at the molecular level. These great advances have brought with them some breath-taking insights into the molecular mechanisms used by nature for replicating, distributing and modifying biological information. We have learned a great deal about the chemical and physical nature of the macromolecular nucleic acids and proteins, and the manner in which carbohydrates, lipids and smaller molecules work together to provide the molecular setting of living systems. It might be said that these few decades have replaced a near vacuum of information with a very large surplus.

It is in the context of this flood of information that this series of monographs on molecular biology has been organized. The idea is to bring together in one place, between the covers of one book, a concise assessment of the state of the subject in a well-defined field. This will enable the reader to get a sense of historical perspective—what is known about the field today—and a description of the frontiers of research where our knowledge is increasing steadily. These monographs are designed to educate, perhaps to entertain, certainly to provide perspective on the growth and development of a field of science which has now come to occupy a central place in all biological studies.

The information in this series has value in several perspectives. It provides for a growth in our fundamental understanding of nature and the manner in which living processes utilize chemical materials to carry out a variety of activities. This information is also used in more applied areas. It promises to have a significant impact in the biomedical field where an understanding of disease processes at the molecular level may be the capstone which ultimately holds together the arch of clinical research and medical therapy. More recently in the field of biotechnology, there is another type of growth in which this science can be used with many practical consequences and benefit in a variety of fields ranging from agriculture and chemical manufacture to the production of scarce biological compounds for a variety of applications.

This field of science is young in years, but it has already become a mature discipline. These monographs are meant to clarify segments of this field for the readers.

Cambridge, Massachusetts   Alexander Rich
                                                             Series Editor

Preface

Some thirty years ago Chargaff (1955) wrote in a review that "the bulk of the nitrogenous compounds of all deoxypentose nucleic acids (*sic*) is composed of two purines, adenine and guanine, and two pyrimidines, cytosine and thymine. The principal exceptions are the deoxypentose nucleic acids of the bacteriophages T2, T4, and T6 of *E. coli* in which 5-hydroxymethylcytosine takes the place of cytosine, and the nucleic acid of wheat germ, in which about one quarter of the cytosine is replaced by 5-methylcytosine. The latter has been recognized as occurring in traces in several deoxypentose nucleic acids of animal origin." It was, however, at least another five years before any real efforts were directed toward assessing the role of 5-methylcytosines in DNA. Indeed they were for a while dismissed as having no biological significance, possibly representing some vestigial remnant of an earlier functional component or, worse still, simply as repositories of excess methyl groups. Nevertheless while Chargaff correctly surmized that 5-methylcytosines could take the place of cytosines in the DNA double helix, he pointed out that this was clearly not occurring at random. The 5-methylcytosine content of the DNAs of a given species was remarkably constant. However what he found "more disturbing" was "that 5-methylcytosine is distributed unevenly in the fractions obtained by the fractionation of calf thymus deoxyribonucleic acid."

Rather than *disturbing* molecular biologists, that sort of observation has been very much a catalytic factor in stimulating the very large research effort now aimed at understanding the significance of the nonrandom occurrence of 5-methylcytosine in DNA. The possibility that the arrangement of methylcytosines in DNA may underpin the mechanisms that regulate gene expression in eukaryotes has recently captured the imagination of many biologists. While this is still very much an open question, methylation of DNA bases (adenine as well as cytosine) became biologically respectable with the discovery of its vital role in the bacterial modification-restriction phenomenon, with its spectacular biotechnological spin-off. In spite of this however, it is still fair to say that the study of DNA methylation continues to provide more questions than it has answered.

While there is still a great deal yet to be learned about DNA methylation, there has been an impressive growth over the last few years in our basic knowledge of the phenomenon. Much effort involving sophisticated chemical analyses and the elegant approaches of modern molecular biology has permitted the rigorous testing of various hypotheses regarding function. For this reason it is perhaps timely to bring together in a single text a relatively unbiased description of the present state of knowledge. Thus we include the doubts and the certainties as well as the challenges that still have to be met if we are to understand the whole biological significance of DNA methylation.

Roger L.P. Adams
Roy H. Burdon

Acknowledgments

We would like to thank all our colleagues with whom we have discussed the problems of DNA methylation and in particular those who have given permission for us to reproduce their published or unpublished work. We are grateful to the authors and publishers of the following figures for their permission to reproduce them in this book.

Figure 2.3	Woodcock, Adams, and Cooper	(1982)
2.4	Gruenbaum, Szyf, Cedar, and Razin	(1983)
3.4	Brown and Dawid	(1968)
5.1	Levy and Walker	(1981)
5.2	Rubin and Modrich	(1977)
5.4	Burckhardt, Weisemann, and Yuan	(1981)
5.12	Wang, Huang, and Ehrlich	(1984)
6.4	Zyskind and Smith	(1980)
7.2	Pech, Streeck, and Zachau	(1979)
7.4	Pollack, Kasir, Shemer, Metzger, and Szyf	(1984)
7.9	Bird	(1980)
7.11	Bird, Taggart, and Smith	(1979)
7.12	Singer, Roberts-Ems, and Riggs	(1979a)
8.4	Taylor and Jones	(1979)
8.5	Kruczek and Doerfler	(1982)
8.8	Rogers and Wall	(1981)
8.9	Meijlink, Philipsen, Gruber, and Geert	(1983)
8.11	Wilks, Seldran, and Jost	(1984)
8.12	Ott, Sperling, Cassio, Levilliers, Sala-Trepat, and Weiss	(1982)
8.13	Busslinger, Hurst, and Flavell	(1983b)
13.1	Holliday and Pugh	(1975)

We are also very grateful to the secretarial staff of the Glasgow University, Department of Biochemistry, and in particular Mrs. Sadie Brown, for their patience while we were writing and rewriting this book. Our thanks are also due to the Medical Illustration Department at Glasgow University for the production of the illustrations under the very able direction of Mr. Ian Ramsden.

Contents

1

Introduction

1.1 Introducing DNA Methylation

The sequence of nucleotides in DNA dictates the sequence of amino acids in the proteins of a cell and hence the form and function of a cell. This is the so-called central dogma of molecular biology. Because of the sequence complementarity of the strands of the DNA duplex, this sequence can be transferred to daughter cells following semiconservative replication and cell division. The replication of DNA occurs with high fidelity and, in general, only one round of replication occurs per cell cycle. Any errors that arise are corrected and any damage to DNA is usually quickly repaired. The DNA in the cell is neither naked, nor is it stretched out in the familiar B form revealed by fiber X-ray diffraction studies. Rather it is associated with proteins and RNA in such a way that it can be packed into the cell and yet transcribed and replicated in an orderly manner.

Traditionally it was believed that DNA was made up of nucleotides containing the four bases, adenine, cytosine, guanine, and thymine, and that the sequence of these bases and the arrangement of these sequences relative to each other was sufficient to distinguish one organism from another. *DNA methylation* that occurs *in vivo* as a result of enzymic action is a specific postsynthetic modification of DNA involving the covalent attachment of methyl groups to certain of the above-mentioned major bases. Such a phenomenon offers a means of altering the base sequence in specific regions of DNA in a regulated and permanent manner.

The uses made by cells of this epigenetic alteration of the DNA sequence are not yet fully appreciated. Nevertheless some of the uses have been established by firm evidence, while others remain speculative and matters of considerable controversy. For instance, while the presence of these naturally added methyl groups on specific bases does not appear to alter the basic rules of complementary base pairing they can bring about changes to the physical structure of the DNA that may be subtle or extensive (see Chapter 3). Alternatively, or in addition, the effect of such base methylation may be to alter the affinity of specific proteins for particular nucleotide sequences present in the DNA. This is clearly established as the basic situation in the operation of bacterial restriction-modifications systems (Chapter 6) and possibly in the phenomenon of maternal inheritance of chloroplast DNA (Chapter 11). On the other hand, there is

Table 1.1. Some Recent Reviews on DNA Methylation

Holliday & Pugh (1975)	DNA modification mechanisms and gene activity during development.
Sager & Kitchin (1975)	Selective silencing of eukaryotic DNA.
Taylor (1979)	Enzymatic methylation of DNA; patterns and possible regulatory roles.
Drahovsky & Boehm (1980)	Enzymatic DNA methylation in higher eukaryotes.
Burdon & Adams (1980)	Eukaryotic DNA methylation.
Razin & Riggs (1980)	DNA methylation and gene function.
Razin & Friedman (1981)	DNA methylation and its possible biological roles.
Doerfler (1981)	DNA methylation, a regulatory signal in eukaryotic gene expression.
Ehrlich & Wang (1981)	5-methylcytosine in eukaryotic DNA.
Felsenfeld & McGhee (1982)	Methylation and gene control.
Cedar et al. (1983)	Effect of DNA methylation on gene expression.
Adams & Burdon (1983)	DNA methylation in eukaryotes.
Doerfler (1983)	DNA methylation and gene activity.
Cooper (1983)	Eukaryotic DNA methylation.
Bird (1984)	DNA methylation: How important in gene control?
Jaenisch & Jahner (1984)	Methylation, expression, and chromosomal position of genes in mammals.
Trautner (ed.) (1984)	Methylation of DNA.
Adams & Burdon (1982)	DNA methylases.
Meselson et al. (1972)	Restriction and modification of DNA.
Bingham & Atkinson (1978)	Restriction endonucleases and modification methylases in bacteria.
Yuan (1981)	Structure and mechanism of multifunctional restriction endonucleases.
Modrich & Roberts (1982)	Type II restriction and modification enzymes
Taylor (1984)	DNA methylation and cellular differentiation
Razin, A., Cedar, H. & Riggs, H.D. (1984)	DNA methylation: biochemistry and biological significance

still speculation as to whether such an explanation might underpin the regulation of DNA transcription (see Chapters 7–10) or be involved in the initiation of DNA replication, repair, or recombination events (Chapters 6, 7, and 12).

If every time a particular DNA nucleotide sequence occurs it is methylated, then no further sequence information is available. For such methylations to have a function, the critical sequence must be present in both a modified and an unmodified form. The unmethylated form may exist only transiently following replication or may exist only in foreign DNA introduced into a cell following infection or mating. In multicellular organisms a particular nucleotide sequence may be methylated in some cells and not in others, and this may affect the interaction of regulator proteins.

In recent years, evergrowing numbers of research reports have suggested several different roles for DNA methylation, and we expect that this flood of information will continue. Of course DNA methylation may have multiple functions within a single cell, and certainly any recent all-embracing theories have tended to struggle against a flood of sometimes inexplicable or contradictory findings. Nonetheless there is already such a wealth of information available that it is timely that it should be gathered together to develop a comprehensive picture.

A number of reviews on the subject have appeared and some pertinent ones are listed in Table 1.1

1.2 Methyl Bases in DNA

In addition to the four major bases present in DNA, a small proportion of methylated bases may also be present. Their structures are given in Figure 1.1. The two that concern us primarily in this book are 5-methylcytosine and N^6-methyladenine.

5-Methylcytosine is in many ways analogous to thymine in structure, but 6-methyladenine has the methyl substituent in the amino group, which might be expected significantly to alter the properties of the base and nucleotide (see Chapter 3).

N^4-methylcytosine is the product of DNA modification by the Bcn I methylase (Janulaitis *et al.*, 1983), and the *mom* gene of phage mu is responsible for a complex modification of adenine residues in infected cells (see Section 6.3). A modified cytosine has also been detected in trypanosome DNA. This is not 5-methylcytosine, but its chemical nature has not yet been elucidated (Raibaud *et al.*, 1983; Pays *et al.*, 1984).

While the level of methylated bases is usually a small fraction of the level of the major nonmethylated bases, the situation in the DNA of the phage Xp12 is quite distinct. In this unusual DNA the base cytosine is *completely* replaced by 5-methylcytosine. In the DNA of the T-even bac-

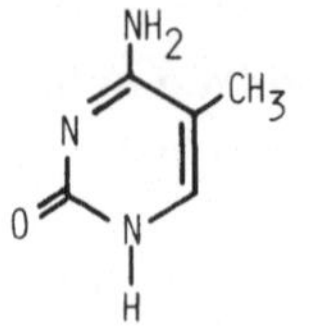

5 methylcytosine

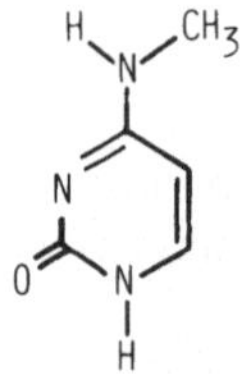

N^4- methylcytosine

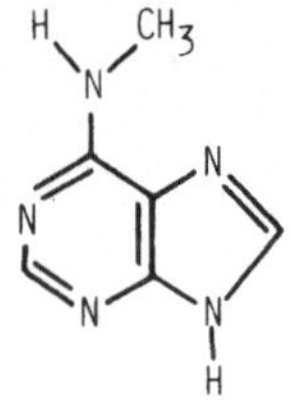

N^6- methyladenine

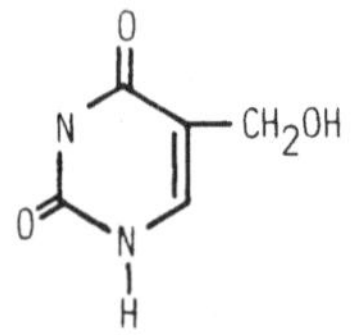

5-hydroxymethyluracil

Figure 1.1. Minor bases found in DNA.

teriophages, cytosine is again *completely* replaced, but this time by hydroxymethylcytosine (which in turn can also be glycosylated). Similarly in the DNA of various *B. subtilis* phages, the thymine can be completely replaced with uracil or hydroxymethyluracil (Takahashi & Marmur, 1963; Kallen *et al.*, 1962). When complete replacement of one of the normal bases occurs it should be emphasized that this is not the result of a postsynthetic modification to the DNA as is the case with the minor methylated bases, but rather the DNA polymerase incorporates the unusual nucleotide that replaces the normal analogue in the precursor pool (see Section 2.1). In some cases postsynthetic modification may occur to DNA that already has one of the normal bases completely replaced by an unusual base.

1.3 Distribution of Methylated Bases

Table 1.2 gives the amount of minor bases in DNA from a variety of species; more details can be found in Chapter 11. Some species have both

5-methylcytosine and 6-methyladenine in their DNA, others have neither. 6-methyladenine is more common in prokaryotes and in lower eukaryotes. In eukaryotes the amount of 5-methylcytosine varies from about 6 mole percent in some plants, *i.e.*, about 30% of cytosines are methylated, to very low or scarcely detectable levels in yeast and *Drosophila*. Even in these latter cases there may be up to 1000 methylcytosines per haploid genome. Although at the limits of detectability by the methods presently in use, this low number still represents approximately one methylcytosine per gene.

There are two problems with the chemical detection of such small amounts of methylcytosine. First, one must be certain that the minor base is in DNA and not in contaminating RNA, and second, one must be certain that the tissue investigated is not contaminated with foreign bacteria or mycoplasma. However, the level of infection with mycoplasma to account for a 6-methyladenine content of 0.03 mole percent would require that 5% or more of the DNA analyzed came from the contaminating organism. This would have a gross effect on growth and be readily detected by the standard screening techniques in use in most cell culture laboratories (Adams, 1980).

The presence of significant but variable levels of methylcytosine in mealy bugs and its virtual absence from some other insects need not be cause for concern. A similar situation occurs between *E. coli* B and *E. coli* C, showing that in the former the function of methylcytosine is obviously dispensable. It is probable that different species put methyl bases to different uses and in their absence find alternative ways of solving common problems.

The amount of methylcytosine in mitochondria and chloroplasts is also a matter of some debate. The use of some methods fails to demonstrate any methylcytosine (*e.g.*, Groot & Kroon, 1979) yet other workers have found a higher proportion of cytosines methylated in mitochondrial DNA than in the corresponding nuclear DNA (Vanyushin & Kirnos, 1974, 1977). These latter results are difficult to reconcile with those of other workers. Sensitive methods have recently confirmed the presence of low levels of methylcytosine in the mitochondrial DNA of mouse cells. Interestingly, methylation of chloroplast DNA in *Chlamydomonas* is normally restricted to female gametes, but the *me*-1 mutation leads to additional extensive methylation of chloroplast DNA in both male and female vegetative cells (Sager, *et al.*, 1984).

In animals, methylcytosine is found predominantly in the sequence -CG- (see Section 7.1). In plants, additional methylations occur apparently in the sequence -CNG-, where N can be any nucleotide. In bacteria, methylcytosine and methyladenine have been found in a wide variety of sequences that are usually but certainly not always symmetrical (see Chapter 6). Two or more separate modifications may be present in a single

Table 1.2. Amount of Minor Bases in the DNA of Various Organisms[a]

Species	Methylcytosine (% of all cytosine)	Methyladenine (% of all adenine)	Reference
Prokaryotes			
E. coli B	—	2.1	Vanyushin et al., 1968
E. coli C	1.0	2.1	Vanyushin et al., 1968
M. luteus	0	0	Vanyushin et al., 1968
Micoplasma arginini	0	2.0	Razin & Razin, 1980
Acholeplasma laidlawii K2	2.3	0.3	Dybvig et al., 1982
Lower eukaryotes			
Tetrahymena pyriformis	<0.03	0.6	Hattman et al., 1978
Tetrahymena thermophila	<0.01	0.8	Pratt & Hattman, 1981
Chlamydomonas reinhardi	0.7	0.9	Pratt & Hattman, 1981
Saccharomyces cerevisiae	<0.03	0	Proffitt et al., 1984
Chlorella vulgaris	11.6	—	Pakhomova et al., 1968
Pseudococcus spp	0.4–1.3	0	Scarbrough et al., 1984
Aedes albopictus	0.17	0.1	Adams et al., 1979b
Locusta migratoria	0.95	—	Wyatt, 1951
Drosophila	0.03	—	Achwal et al., 1984
	<0.02	—	Pollack et al., 1984
Animals			
sea urchin	6.5	ND	Chargaff et al., 1952
	2.9	ND	Vanyushin et al., 1970
salmon	9.1	ND	Vanyushin et al., 1970
herring	8.9	ND	Vanyushin et al., 1973b
frog	7.2	ND	Dawid, 1965
chick	4.5	—	Vanyushin et al., 1973b
	3.9	—	Kappler, 1971
mouse	3.8	—	Pollack et al., 1984
calf thymus	5.4	—	Vanyushin et al., 1970
bull sperm	2.5	—	Adams et al., 1983a
snail	2.9	ND	Vanyushin et al., 1970

			Reference
frog	—	11.6	Vanyushin et al., 1973b
rat	—	4.3	Vanyushin et al., 1973b
pigeon	—	4.6	Vanyushin et al., 1973b
Plants			
wheat	ND	31	Thomas & Sherratt, 1956
bracken	ND	33	Thomas & Sherratt, 1956
broad bean	—	23	Baxter & Kirk, 1969
bluebell	—	~40	Vanyushin & Milner, 1965
pine	—	19.7	Vanyushin & Belozerski, 1959
corn	—	29.3	Vanyushin & Belozerski, 1959
Viruses			
frog virus 3	0	20	Willis & Granoff, 1980
herpes simplex virus	—	0	Low et al., 1969
SV40	—	0	
Epstein-Barr virus	—	14.6	Diala & Hoffman, 1983
Mitochondrial DNA			
frog	ND	0	Dawid, 1974
human	ND	0	Dawid, 1974
yeast	ND	0	Groot & Kroon, 1979
N. crassa	ND	0	Groot & Kroon, 1979
cow	ND	0	Groot & Kroon, 1979
rat	ND	0	Groot & Kroon, 1979
mouse	—	0.3	Pollack et al., 1984
mouse	—	0.2	Nass, 1973
hamster	—	0.5	Nass, 1973
Chloroplast DNA			
vegetative *Chlamydomonas*	—	0	Royer & Sager, 1979
Chlamydomonas female gametes	—	32	Sager et al., 1984
Euglena	—	0	Ray & Hanawalt, 1964
Tobacco	—	ND	Whitfield & Spencer, 1968
Spinach	—	ND	Whitfield & Spencer, 1968

a See Fasman (1976) for more figures.
(ND) not detected
(—) not looked for

bacterium where it is clear that each modification is the result of the action of a distinct enzyme. The DNA-cytosine methyl-transferases of different animal species are, however, closely related structurally and functionally (Chapter 5). In plant cells, it is possible that two methyltransferases exist—one similar to the animal enzyme, methylating cytosines in CpG dinucleotides, and a second acting on CpNpG trinucleotides.

2

DNA Methylation in the Cell

2.1 The Synthetic Reaction

DNA methylation involves the enzyme-catalyzed transfer of a methyl group from S-adenosyl-L-methionine (AdoMet) to either the 5 position on the cyosine ring or the 6-amino group of adenine in polymeric DNA (Figure 2.1) (Borek & Srinivasan, 1966). The latter reaction occurs predominantly in prokaryotes and lower eukaryotes. Methylation of cytosine probably involves intermediate formation of an enzyme-DNA covalent link (see Figure 5.18). This conclusion is reached by analogy with other enzymes that catalyze electrophilic substitutions at the 5 position of pyrimidine rings (Santi *et al.*, 1983). Although free methyldeoxycytidine has been detected in animal tissues (Soska & Bezdek, 1962), it probably arises there by breakdown of DNA. There is no evidence for incorporation of exogenous methyldeoxycytidine into animal cell DNA (Adams *et al.*, 1982), but, following deamination at the nucleoside level, it can substitute for thymidine to support the growth of a thymine-requiring strain of *E. coli* (Cohen & Barner, 1957).

5-Methyldeoxycytidylate on the other hand is synthesized in *Xanthomonas* infected with phage Xp12 and is incorporated directly into the phage DNA in place of deoxycytidylate (Kuo & Tu, 1976). (It is synthesized not from methionine, but using the 3-carbon of serine in a reaction involving tetrahydrofolic acid.) In this context, it perhaps should be pointed out that *E. coli* DNA polymerase can also incorporate 5mdCTP into DNA in place of dCTP (Stein *et al.*, 1983), but the fact that this reaction does not occur *in vivo* must imply that the enzymes required to make the methyl analogue of dCTP are absent from normal cells.

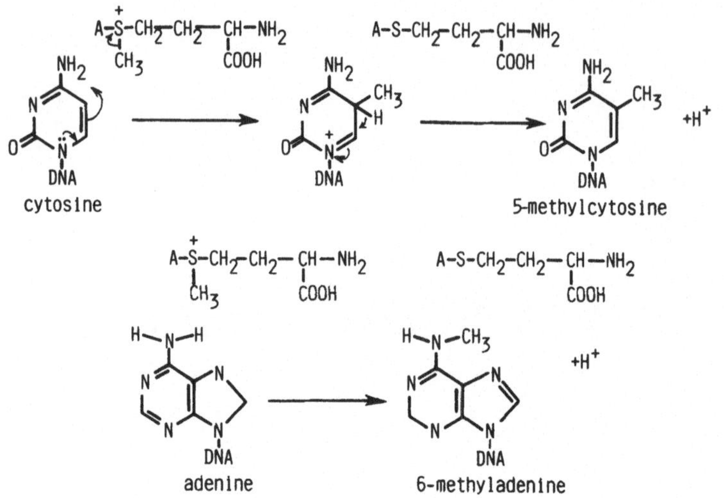

Figure 2.1. Synthesis of the minor methylated bases found in DNA.

2.2a Timing of Methylation in Prokaryotes

All methylation occurs on the nascent DNA daughter strand close to the growing fork (Billen, 1968; Szyf *et al.*, 1982). This is consistent with the finding that the DNA is fully modified and not susceptible to cognate restriction enzymes (see Chapter 6). When the amount of replicating DNA per cell increases (for example, following the induction of a lambda lysogen) the level of DNA methylase is insufficient and the phage DNA is only partly methylated (Szyf *et al.*, 1984). Normally, however, methylation by the *dam* methylase (see Chapter 5) must lag a little behind the replication fork, and in a ligase-deficient mutant, Okazaki pieces accumulate that are deficient in 6-methyladenine (Marinus, 1976). This delay provides the basis of the mismatch repair reaction considered in Section 6.2.2.

DNA deficient in methyl groups accumulates when methionine auxotrophs are grown in the absence of methionine. This treatment also has the effect of preventing the initiation of new rounds of DNA replication, but whether the two events are connected is not certain (see Section 6). On readdition of methionine, the unmethylated DNA is rapidly methylated, and a new round of replication is initiated (Lark, 1968a, 1968b).

2.2b Timing of Methylation in Eukaryotes

Broadly speaking, methylation of DNA in higher eukaryotes takes place in the DNA synthesis (S) phase of the cell cycle (Burdon & Adams, 1969);

Adams, 1971; Lutz *et al.*, 1973; Geraci *et al.*, 1974; Turkinton & Spielvo-gel, 1971; Bugler *et al.*, 1980; Woodcock *et al.*, 1982). However, complete methylation of DNA is not achieved for several hours after DNA replication, and this lag is accentuated when cells are grown in methionine-deficient medium (Culp & Black, 1971; Kiryanov *et al.*, 1982; Adams *et al.*, 1974). Furthermore, methylation is less sensitive to inhibitors of DNA synthesis than is DNA replication. On addition of hydroxyurea or ami-nopterin to cells, methylation continues on DNA made before the inhibitors were added (Figure 2.2). When cells are incubated simultaneously with [^{3}H-Me]methionione and bromodeoxyuridine for a 1- to 6-hour period prior to harvesting, 10 to 15% of the methyl groups are found to band with light/light DNA on neutral cesium chloride gradients, showing that they were incorporated into DNA made prior to the addition of the radio-activity (Figure 2.3). This lag is most conspicuous for DNA made late in S phase (Adams, 1971). The methyl groups in the light/heavy peak can be shown by alkaline buoyant density centrifugation to be 100% in the heavy, daughter strand (Bird, 1978; Woodcock *et al.*, 1982), and there is no evidence from these experiments that methylation is taking place anywhere except on the newly synthesized daughter strand of DNA. The conclusion is that methylation is complete before a new round of replication is initiated.

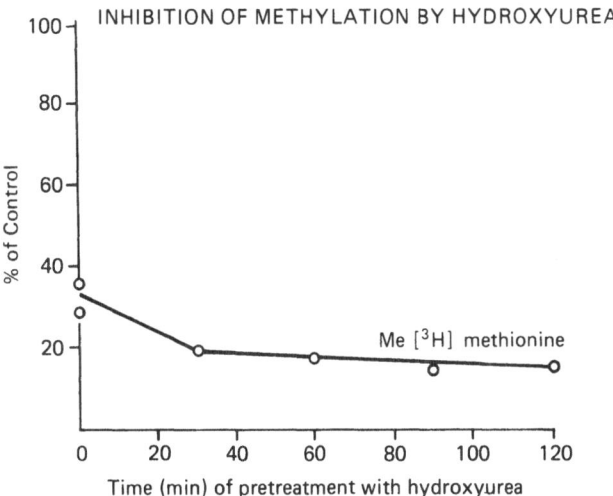

Figure 2.2. Inhibition of DNA synthesis and methylation by hydroxyurea. Mouse cells were preincubated with 2 mM hydroxyurea prior to labeling with ^{3}H thymidine (to measure DNA synthesis) or Me[^{3}H] methionine under conditions where only methylcytosine in DNA becomes labeled. Although synthesis is immediately inhibited by 95%, methylation continues at about 15% of the control level for up to 2 hours.

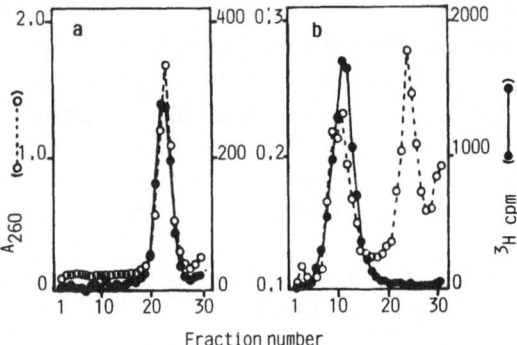

Figure 2.3. Delayed methylation in human cells. Cells were incubated with bromodeoxyuridine to render newly synthesized DNA strands heavy, and with Me[³H] methionine to label methylcytosine. The light/heavy and light/light duplex DNA fractions were separated by two cycles of centrifugation in neutral CsCl gradients. The purified light/light (a) and light/heavy (b) DNA fractions were banded in alkaline CsCl/Cs₂SO₄ gradients. Panel (a) illustrates the presence of methyl groups that were added to DNA synthesized prior to addition of the labels. (From Woodcock *et al.*, 1982.)

In contrast to these experiments, Kappler (1970) labeled cells with ¹⁴C deoxycytidine and measured the rate of incorporation of radioactivity into DNA cytosine and methylcytosine. He found that incorporation increased linearly with time and, on extrapolating backwards, he concluded that methylation is complete within 1.6 minutes of synthesis of the DNA. The amount of radioactivity incorporated in these experiments is, however, very low. Nevertheless, Gruenbaum *et al.* (1983), using a different technique, also claim that the lag between synthesis and methylation is no more than 1 minute. They permealized mouse L929 cells by treatment with hypotonic buffer and introduced [α³²P]dGTP into them for 45 seconds before replacing the growth medium. At various times later, cells were harvested, DNA prepared and digested to the 3′ mononucleotides, which were separated by two-dimensional chromatography. The extent of methylation of CpG dinucleotides was calculated from the radioactivity in methyl dCMP and dCMP (Figure 2.4). Methylation could be inhibited by carrying out the permealization in the presence of S-adenosylhomocysteine. The percent of methylation achieved in this experiment was 65% of the CpGs, which corresponds approximately to the level found in mouse cells. Despite these findings relating to short-term kinetics of DNA methylation, the proportion of cytosines methylated within 50 minutes of synthesis of DNA is smaller than that found in mature DNA. By labeling mouse cells with 6³H uridine for different times and then analyzing the DNA bases by HPLC, the level of methylation was found to rise from 2.5 to 4% over a 2-day period (Adams *et al.*, 1984; Davis *et al.*, 1985).

Perhaps there is no conflict between the two approaches. One is dealing with very short times and claims that after a very short lag of one or two

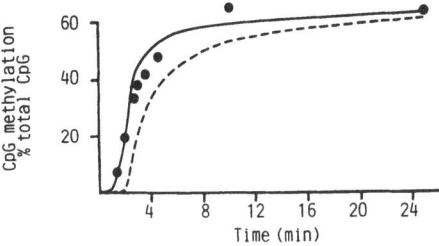 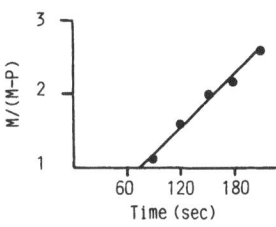

Figure 2.4. Kinetics of DNA methylation in permealized mouse cells. Cells were treated for 45 seconds with hypertonic buffer containing $\alpha[^{32}P]dGTP$, and then fresh medium was added. At the indicated times DNA was isolated and hydrolyzed to the 3' monophosphates, which were separated by 2D TLC (see Figure 7.4). The radioactivity in dCMP and methyl dCMP gives the proportion of CpG dinucleotides methylated. Theoretical curves for a 1-minute lag (solid) and a 2-minute lag (dashed) were obtained by plotting M(T-X)T vs time where M is maximal methylation (taken as 66%), T is the labeling time, and X the assumed lag time. (From Gruenbaum *et al.*, 1983.)

minutes methylation is complete. The other, dealing with much longer times, leads to the conclusion that after the initial rapid methylation there is a period of several hours when DNA replication is over, during which additional methyl groups are added to the DNA. This *delayed methylation* in the apparent absence of DNA synthesis will be considered further below.

The lag of about a minute between synthesis and methylation is consistent with the finding that Okazaki pieces are normally ligated prior to methylation (Adams, 1974; Hotta & Hecht, 1971; Drahovsky & Walker, 1975). The possible effects of chromosomal proteins on the methylation process will be considered in Chapter 12.

Although the lag between synthesis and complete methylation is normally only a few hours for mammalian DNA, there is evidence that methylation may only be completed during mitosis (Bugler *et al.*, 1980). Similar results have been found in *Lilium* (Hotta & Hecht, 1971), and in *Physarum polycephalum* methylation of DNA continues for six or seven mitotic cycles following the S phase in which the DNA is made (Evans & Evans, 1970; Evans *et al.*, 1973). In sea urchins, that DNA made immediately after fertilization of the egg continues to accept methyl groups throughout several cell division cycles (Adams, 1973).

2.3   *De Novo* and Maintenance Methylation

In eukaryotes methylcytosine is found predominantly if not exclusively in the dinucleotide mCpG (Section 7.1). Following replication an unmethylated CpG in the daughter strand of DNA will be paired with the mCpG in the parental strand. Methylation of such newly synthesized CpGs in hemimethylated duplex DNA will maintain the pattern of methylation from one generation to the next (Figure 2.5). Such hemimethylated sites are the preferred substrate for purified DNA methylases (Section 8.2.5). The fact that, *in vivo,* methylation occurs only on the newly synthesized daughter strand of DNA when paired unmethylated CpGs occurring in other regions of DNA are not methylated is a further argument that the methylcytosine in the parental strand forms part of the recognition signal for the enzyme.

Bird (1978) has shown using restriction enzymes that in the ribosomal DNA of *Xenopus laevis* significant numbers of hemimethylated sites cannot be detected (Figure 2.6). Thus all CpGs are either methylated on both strands of the DNA or not methylated on either strand. (See Chapter 6 for a discussion of restriction enzymes.)

That the pattern of methylation is maintained from one generation to the next is amply demonstrated by transfection experiments such as that carried out by Wigler *et al.* (1981) where a cloned chicken thymidine kinase gene was methylated at CCGG sequences with Hpa II methylase and introduced into thymidine kinaseless mouse cells. Twenty-five generations later the state of methylation of the Hpa II sites was shown to have been maintained with just less than 100% fidelity (see Section 8.4.2).

If the consequence of a high-fidelity DNA methylase is to maintain the pattern of methylation, how is that pattern established in the first place,

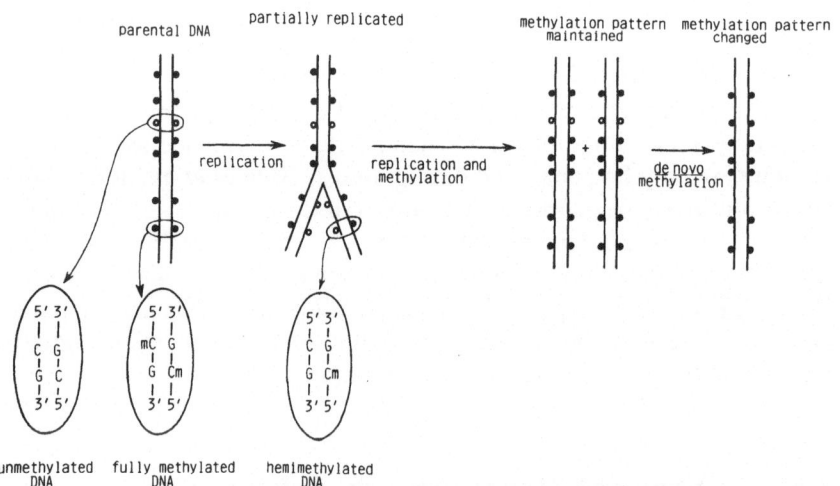

Figure 2.5. Maintenance of the methylation pattern at DNA replication and *de novo* methylation.

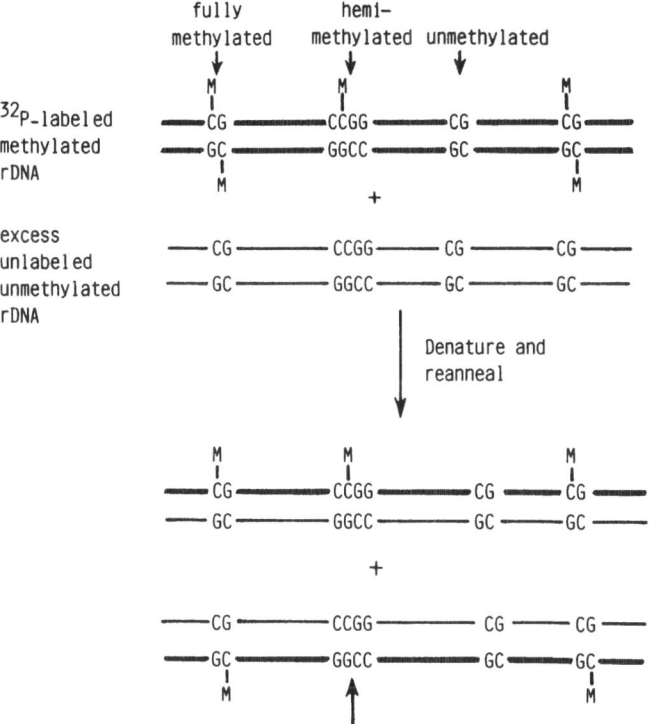

Figure 2.6. Demonstration of the absence of hemimethylated sites in ribosomal DNA. The heavy lines indicate the presence of ^{32}P in the cellular DNA, and possible fully methylated, hemimethylated, and unmethylated sites are indicated. Following annealing with an excess of unlabeled, unmethylated, ribosomal DNA, the DNA is treated with the restriction enzyme HpaII. An increase in cleavage relative to the original DNA would be evidence for the presence of hemimethylated sites.

and how can it be changed to produce the different patterns found in different tissues (Section 8.1)? It is to find the answer to these questions that many researchers have recently become particularly interested in investigating eukaryote DNA methylation.

Although somatic cells appear to have little ability to add methyl groups to regions of DNA previously devoid of methylcytosine, there is an activity present in very early mouse embryos (but not in *Xenopus* embryos) that will carry out such *de novo* methylation of injected DNA and also of endogenous unmethylated satellite DNA (see Section 8.2). This *de novo* methylation may occur to all DNA that is not present in active regions of chromatin and may represent a mechanism whereby the developmental clock may be reset in the very early embryo (see Figure 2.5).

2.4   Demethylation

Not only is the pattern of methylation of particular genes stably maintained from generation to generation, but also methyl groups, once added to DNA, appear as stable as the DNA itself (Burdon & Adams, 1969; Evans *et al.*, 1973). Furthermore the demethylation associated with the transcriptional activation of genes (Section 8.3) takes two cell cycles or more to become apparent and can be explained by DNA replication occurring in the absence of methylation (Figure 2.7). On that basis it seems unlikely that an enzyme exists that can remove methyl groups from methylcytosine in DNA. Indeed the major breakdown pathway involves release of the pyrimidine from the DNA, followed by ring cleavage (Adams *et al.*, 1981a). Certainly no enzyme exists to remove the methyl group from thymine. However, Gjerset and Martin (1982) have reported the presence in Friend cells of an enzyme capable of demethylating polymeric DNA. The activity was low, and Adams *et al.* (unpublished data) have not

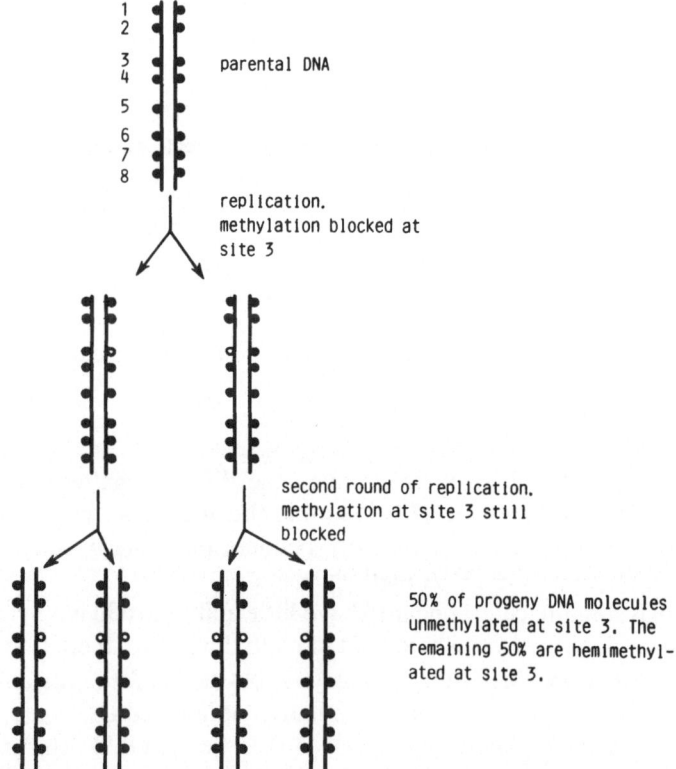

Figure 2.7. Regions or sites on DNA can be demethylated following two rounds of replication in the absence of methylation.

been able to obtain convincing evidence for its presence in a variety of other tissues.

Nevertheless it must be conceded that there are cases when demethylation occurs in the absence of substantial DNA replication (*e.g.*, Wilks *et al.*, 1984). Although this could result from the action of a site-directed demethylase, we suggest in Chapter 12 that it may be the result of repair of DNA in the absence of methylation. Such repair may be more likely in the exposed regions, which are present near the 5' ends of genes (Adams *et al.*, unpublished data).

2.5   Methylation in Isolated Nuclei

When nuclei are isolated from mammalin cells or from sea urchins and incubated with [^{3}HMe]S-adenosyl methionine, radioactive methyl groups are enzymically transferred to cytosines in the endogenous DNA (Burdon, 1966; Tosi *et al.*, 1972; Adams & Hogarth, 1973; Geraci *et al.*, 1974; Cox *et al.*, 1977; Davis *et al.*, 1985). As the isolated nuclei do not make DNA, this is further evidence that methylation can continue even if DNA replication has ceased. As mentioned above, this is a characteristic of *delayed* methylation. Indeed specific prior treatment of cells for 1 hour with hydroxyurea to inhibit DNA synthesis completely only slightly reduces the ability of the nuclei isolated from such cells to methylate their endogenous DNA. Again this supports the notion that in isolated nuclei we are looking predominantly at *delayed methylation.*

To analyze what type of DNA is being methylated during delayed methylation in isolated nuclei, Adams and Hogarth (1973) incubated mouse L929 cells for increasing times in the presence of bromodeoxyuridine. The cells were harvested together 19 hours after subulture and the nuclei isolated and then incubated for 60 minutes with [^{3}HMe]S-adenosyl methionine. Figure 2.8 shows the distribution of tritium in the DNA samples after centrifugation to equilibrium on gradients of alkaline cesium chloride. When the density label was present for less than 75 minutes before the cells were harvested, the major fraction of methylation occurring in isolated nuclei took place on light DNA, *i.e.*, DNA more than 75 minutes old. However when the density label was present from the time of subculture, all the *in vitro* methylation occurred on heavy DNA, indicating that parental strands are not methylated and that all the methylation in isolated nuclei, therefore, is of the maintenance type and occurs at hemimethylated sites.

How then can delayed methylation be explained? As already mentioned, in the intact cell the initial rapid methylation takes place after the Okazaki pieces are joined and after nucleosomes have formed on the DNA (a process that takes only 15 to 25 seconds) (Weintraub, 1979). It

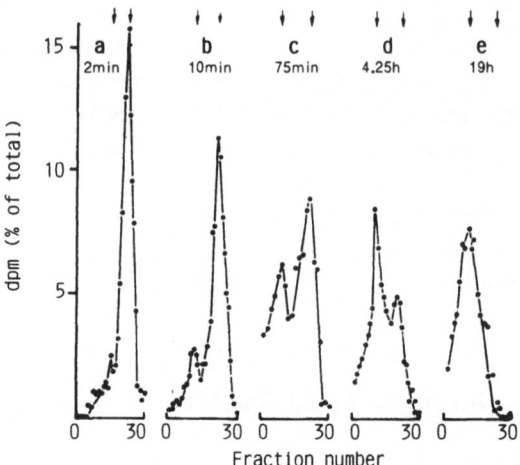

Figure 2.8. Delayed methylation measured in isolated mouse cell nuclei. Cells, subcultured from stationary phase at time zero, were incubated with bromo-deoxyuridine and ^{14}C thymidine for the indicated times and harvested together at 19 hours. Nuclei were prepared and incubated with Me[^{3}H]AdoMet for 60 minutes prior to isolation of the DNA, which was centrifuged to equilibrium on gradients of alkaline CsCl. (From Adams & Hogarth, 1973.)

may be that the nucleosomal proteins have an inhibitory effect on DNA methylation *in vivo*. Their effect may be to render a number of sites inaccessible to the methylating enzyme, which may now have to wait for an opportunity to interact with these obstructed, hemimethylated sites. Such an opportunity may arise within the few hours following replication, perhaps at mitosis, and may explain the origin of the delayed methylation. (However, if the obstruction is more permanent and a second round of replication occurs, then the DNA will become demethylated.) On the other hand (Kiryanov *et al.* (1982) and Vanyushin (1984) suggest that the delayed methylation is caused by methylation of short stretches of DNA made when Okazaki pieces are joined together. This conclusion is consistent with their earlier observation (Kiryanov *et al.,* 1980) that in mouse L cells grown at high cell densities, Okazaki pieces accumulate for 10 minutes and more after synthesis. These accumulated Okazaki pieces have only 2.8% of their cytosines methylated, compared with 4.3% for total DNA. This delay observed in the joining would require that 200 or more Okazaki pieces would accumulate at each growing fork prior to ligation and that these Okazaki pieces are packaged into nucleosomes. As cells grown in the presence of S-isobutyladenosine have only 2.8% of their DNA cytosines methylated, these workers suggest that two different methylating enzymes may be active; one methylates the Okazaki pieces, and a second, which is sensitive to the inhibitor, is responsible for the delayed methylation.

3

DNA Structure and the Effect of Methylation

3.1 Structure of DNA

The traditional B form of DNA is a right-handed antiparallel double helix. It is in this form that DNA exists in solution at physiological salt conditions. It should be borne in mind however that the structure of DNA even in the B form is far from regular, and the helical twist angle can vary between 40.0° and 27.7°, giving DNA molecules with an average of 10.4 to 10.6 base pairs per turn. In this form the base sugar linkage is normally anti with the C8 (purine) or C6 (pyrimidine) above the sugar ring, which has the endo configuration (see Figure 3.1) (Dickerson, 1983a; Wang, 1979; Rhodes & Klug, 1981; Trifonov, 1982).

Where the sequence of the DNA has alternating purine and pyrimidine residues, the more or less smooth B form of DNA takes up a wrinkled conformation in which the orientation of the phosphate linking pyrimidine (5') to purine (3') nucleosides is twisted (Figure 3.2) (Arnott *et al.*, 1983). This wrinkled structure may play an important part in DNA protein interactions, though Dickerson (1983b) points out that this model contains no further sequence information not present in the standard B-DNA conformation.

Under conditions of high salt, DNA regions with alternating purines and pyrimidines undergo a major structural change involving rotation of the purine ring into the syn configuration in which the C2 rather than C8 is above the sugar ring (Figure 3.1). This produces DNA with a left-handed double helix and gives the DNA backbone a zigzag shape. N7 and C8 of guanine and adenine bases are outside the helix in a very exposed position. Normal base stacking can thus no longer occur, but rather the pyrimidines can stack with pyrimidines of the opposite strand (Zimmer-

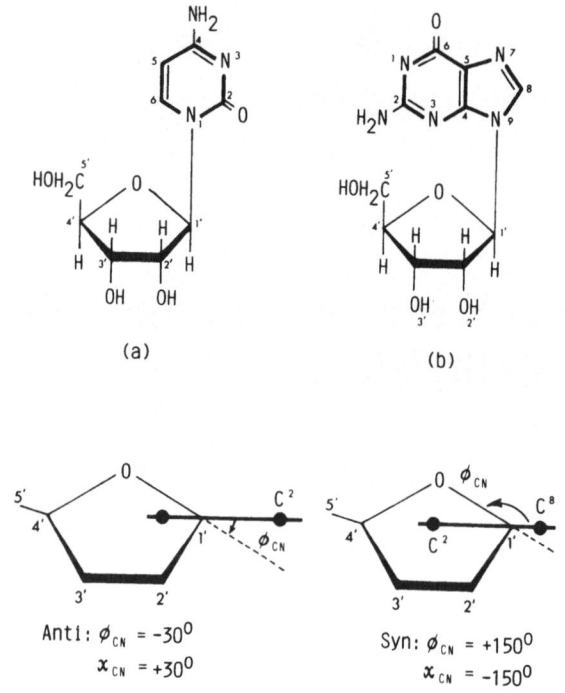

Figure 3.1. The anti and syn configurations of nucleosides.

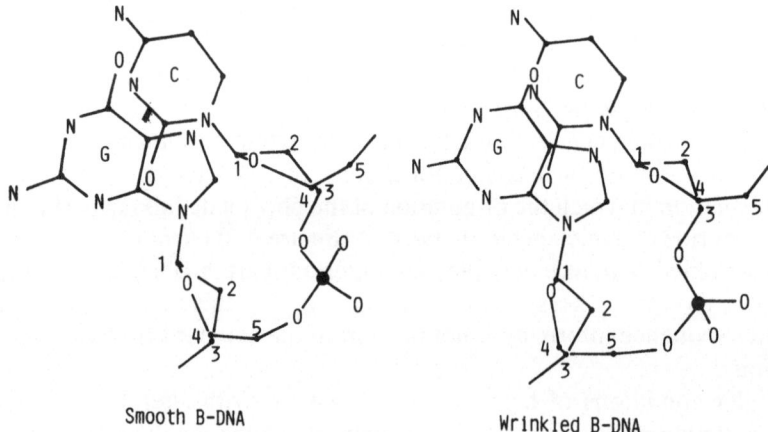

Figure 3.2. The CpG dinucleotide is drawn in two conformers of B-form DNA, the smooth and the wrinkled form.

man, 1982). This left-handed duplex is called Z-DNA, and the ease of formation of Z-DNA depends on a number of factors including base composition. Thus whereas a duplex of alternating cytosines and guanines (poly dC-dG:poly dC-dG) will only form Z-DNA in solution at 2.7 M NaCl or 700 mM MgCl$_2$; the methylated form (poly mdC-dG:poly mdC-dG) will form Z-DNA at much lower salt concentrations (0.7 M NaCl or 0.6 mM MgCl$_2$) which approach physiological conditions (Behe & Felsenfeld, 1981; Pohl & Jovin, 1972).

3.2   Z-DNA and Supercoiling

Formation of Z-DNA is also favored in regions of supercoiled DNA containing alternating purines and pyrimidines. The conversion from the right-handed B conformer to the left-handed Z-DNA releases some of the torsional stress present in the supercoiled molecule and can occur at physiological salt concentrations (Azorin et al., 1983; Peck & Wang, 1983). Methylation of (dC-dG) regions in supercoiled plasmids reduces the number of negative supercoils necessary to stabilize the Z configuration (Klysik et al., 1983). When the topological constraints are very severe, even nonalternating sequences can assume a Z conformation (Pohl et al., 1982).

In order to assess the relative effects of supercoiling, salt concentration, and cytosine methylation, Wells' group has prepared plasmids derived from pBR322 with inserts of (dC-dG)$_{13}$ and (dC-dG)$_{16}$ and longer. Those with inserts longer than 50 bp are unstable and tend to lose the insert on propagation in E. coli. The plasmids with shorter inserts are less supercoiled than expected following isolation from E. coli, implying the presence of segments of left-handed Z-DNA (Klysik et al., 1982; Singleton et al., 1982, 1983). Levels of supercoiling sufficient to induce regions of Z-DNA are present in bacteria, though doubts have been raised as to whether DNA in eukaryotes is supercoiled at all (Sinden et al., 1980; Pettijohn & Pfenninger, 1980; Nordheim & Rich, 1983). Recent experiments, however, have been interpreted as showing the presence of supercoiled DNA in the transcribing fraction of chromatin in Xenopus oocytes (Ryoji & Worcel, 1984). In supercoiled plasmids, salt has the opposite effect to that expected, as the ability of alternating dC-dG sequences to relax supercoils by forming Z-DNA is reduced as the salt concentration is raised to 200 mM NaCl. However, methylation of the cytosines in the insert reduces the amount of supercoiling necessary to bring about formation of left-handed DNA (Azorin et al., 1983; Lilley, 1983; Singleton et al., 1982).

It is important to stress two points. First, the experiments described so far relate to situations where artificial polymers have been studied either alone or following insertion into plasmids. Below we shall review the evidence for the occurrence of Z-DNA in normal, "unengineered" DNA. Second, although DNA containing alternating sequences of any purine and pyrimidine has the potential to take up the Z configuration, our interest lies in sequences containing the dinucleotide CG (repeated or not) and the effect that cytosine methylation has on the chance that these sequences may take up the Z configuration.

3.3   Cytosine Methylation and Z-DNA *in Vivo*

In order to assess the significance *in vivo* of cytosine methylation in the formation of Z-DNA, it is important to decide whether alternating series of cytosines and guanines occur naturally and if so in what sort of position.

Regions of eukaryotic DNA containing alternating purine and pyrimidine bases are not uncommon, for instance there is a sequence of 50 alternating T and G residues in intron IV of the actin gene (Hamada & Kakunaga, 1982). Using this as a probe, over 40,000 copies of this sequence have been demonstrated in the genome of higher eukaryotes. However, a search by similar means for poly(dG-dC) poly(dG-dC) failed to find any sequences in the yeast or calf genome that would hybridize to the CG probe, although some were found in the human and mouse genome (Hamada *et al.*, 1982). On searching published DNA sequences, we have found no *long* stretches of alternating cytosine and guanine sequences, though the sequence CGCGCGCG does occur in the spacer between 18S and 5.8S rRNA genes in *X. borealis*. When methylated, such a sequence may be long enough to take up a Z configuration (Quadrifoglio *et al.*, 1981), though the significance of this is questionable because this octanucleotide is not present in the same position in DNA from the related *X. laevis* (Furlong & Maden, 1983).

Z-DNA is highly immunogenic and Z-specific antibodies have been used to search for Z-DNA in fixed chromosomal preparations from a number of species. Reactive material has been found in dipteran polytene chromosomes, in the macronucleus of *Stylonychia,* and in primate chromosomes. Antibodies have also been used to detect Z-DNA in supercoiled plasmids (Nordheim *et al.*, 1981; Nordheim & Rich, 1983; Nordheim *et al.*, 1982; Arnt-Jovin *et al.*, 1983; Lipps *et al.*, 1983; Viegas-Pequignot *et al.*, 1983; Morgenegg *et al.*, 1983; Thomae *et al.*, 1983).

These studies may be criticized on a number of grounds. (1) Although highly specific, a small degree of cross-reaction may lead to confusion of Z-DNA with B-DNA, which is present in much higher amounts. (2) Binding of DNA to the antibody may bring about a conformational change to Z-DNA in regions that were previously present as B-DNA. (3) The acid fixation of cells and chromosomes involved in antibody studies leads to the removal of histones from DNA. This will produce a torsional strain in the DNA, which may lead to its taking up a Z-DNA conformation. Studies omitting the fixation step fail to show the presence of Z-DNA (Hill & Stollar, 1983).

Bearing in mind these criticisms, researchers have used antibodies to show the presence of Z-DNA in protozoa, insects, mammals, and plants. As the amount of methylcytosine in insects and some protozoa is very low, the relevance of cytosine methylation to formation of Z-DNA remains obscure.

It has been suggested that conversion of a short stretch of DNA to the Z form may have a dramatic effect on chromatin structure and gene expression. Apart from the Z-DNA itself forming a recognition signal for specific proteins that may regulate transcription, the change in superhelical density brought about by changing a stretch of right-handed double helix to left-handed double helix may bring about a loosening of chromatin structure over domains as long as 100 kbp. In addition, the discontinuity of structure at the boundary between regions of B-DNA and regions of Z-DNA may provide an entry point for RNA polymerase or other enzymes involved in gene function. Certainly this particular structure is recognized by S1 nuclease (which digests single-stranded regions of DNA), and plasmids with inserts of 50 bp of Z-DNA are unstable, suggesting the presence of a structure that is possibly recognized by recombination enzymes. Vardimon and Rich (1984) have shown such stretches of Z-DNA to be resistant to both DNA methylase and restriction endonuclease.

Nickol *et al.* (1982) have shown that nucleosomes do not form on Z-DNA. Thus a region of Z-DNA in chromatin would have the effect of (a) exposing a section of nucleosome-free DNA to regulatory proteins, and (b) providing a reference point beyond which nucleosomes would occur at fixed positions (Phasing). Nuclease-sensitive sites are found towards the 5' ends of active genes where regulatory sequences occur, however, it is not yet known whether these may in part be produced as a result of the presence of Z-DNA.

It is tempting to suggest that methylation of DNA may exert a dramatic effect on gene expression by bringing about a change in conformation of DNA. Thus a methylation event may convert a region of DNA from the B form to the Z form and loosen up a domain of chromatin, thus allowing expression of several genes. Alternatively, a single gene may be activated by either allowing access of RNA polymerase directly at the region of Z-DNA or at some downstream position exposed on the outside of a nucleosome by a phasing event. Arguing against this is the observation that transcription of recombinant plasmids with long stretches of poly(dC-dG) adjacent to transcriptional control sequences is not affected by methylation with HhaI methylase (which should methylate every cytosine in such stretches) (Jove *et al.*, 1984). Transcription was studied in an *in vitro* system that may show a decreased susceptibility to changes in control sequences.

Various models interrelating methylation, Z-DNA formation, and transcription are considered in Chapter 12.

3.4 Z-DNA in Vertebrates

Many of the arguments relating methylation of DNA to its increased propensity to form Z-DNA require that sequences of alternating cytosines and guanines of at least eight bases in length occur with a high frequency throughout the genome. If formation of Z-DNA is a common way of regulating gene expression via methylation, then such sequences should occur several thousand times per genome. Although regions of alternating purine and pyrimidine nucleotides do occur in which occasional CG dinucleotides are present (*e.g.,* Karin *et al.,* 1984), the use of an alternating CG probe failed to show the presence of any long runs of this sequence in calf DNA (see above). Indeed, as the vertebrate genome is deficient in the dinucleotide CG, an alternating octanucleotide of that sequence might only be expected to occur once every 2.4×10^8 bases, *i.e.,* in perhaps ten positions in the genome. (Vertebrate DNA has about 20% cytosine and 20% guanine but the CG dinucleotide only occurs one-fifth as frequently as expected, *i.e.,* once every $5 \times 5 \times 5 = 125$ bases. Four such dinucleotides would occur adjacent to each other once every $125 \times 125 \times 125 \times 125 = 2.4 \times 10^8$ bases.)

This argument, however, is misleading, because it is based on two suppositions:

1. Each region of the genome reflects the average base composition. This is not true, as, for example, the human alpha globin gene region is 67% G+C, but the human β-globin gene region is only 40% G+C.
2. The whole of the vertebrate genome is deficient in the CG dinucleotide. It has been pointed out by Adams and Eason (1984) that the most G+C-rich regions of the genome are not deficient in CG dincleotides. This is clear from Figure 3.3, which shows the frequency of occurrence of the CG dinucleotide plotted against the G+C content for 46 vertebrate gene regions selected from the EMBL nucleotide sequence data library II. Also on the graph are two lines that represent the expected frequency of CG dinucleotide based on a random occurrence (upper line) and on an 80% deficiency (the average for the vertebrate genome). Those sequences poor in G+C are the most deficient in CG dinucleotides, whereas those regions rich in G+C are not at all deficient. These nondeficient regions are ribosomal DNA's (mouse and *Xenopus*), human enkaphalin, chick histone H2A, and a repetitive DNA sequence from monkey. The nonvertebrate eukaryotes show the expected fre-

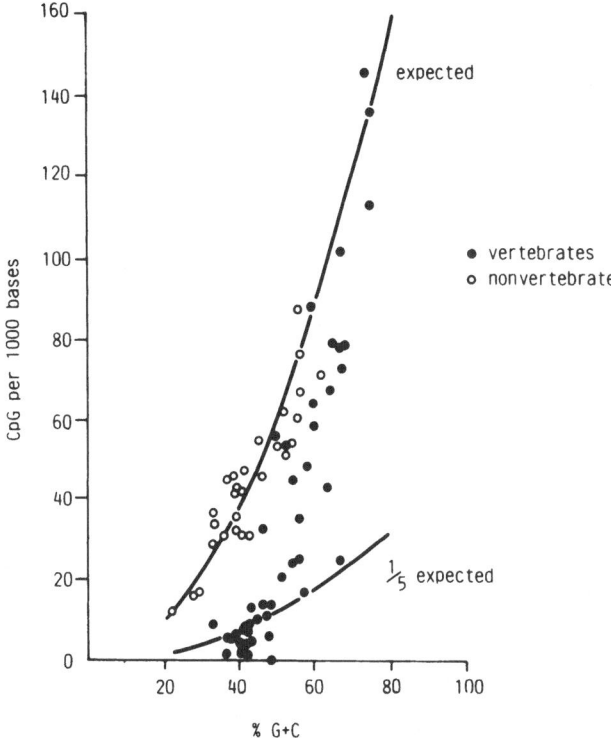

Figure 3.3. The CpG contents of various gene regions selected from the EMBL nucleotide sequence data library are plotted against their G+C contents. The solid lines represent the results expected based on a random distribution of bases and one-fifth of random. ● vertebrates; ○ nonvertebrates (from Adams & Eason, 1984).

quency of occurrence of the dinucleotides. A possible explanation for these observations is that the increased stability of these regions reduces the time they spend in a denatured state and hence reduces the opportunity for such regions to be deaminated (see Section 7.2).

A search of the DNA sequence data library for regions of alternating CG's reveals the figures shown in Table 3.1. It is clear that runs of alternating cytosines and guanines do occur and, in fact, are much more frequent than would be expected on a random basis. The sequence CGCGCGCG has been found five times in vertebrate DNA. It occurs twice as CGCGCGCGCG in the third intron of human gamma globin DNA and in the second intron in mouse histocompatibility *H2LD* gene and about 60 bp upstream from the start of transcription in the human enkaphalin genes. Thus, only in one case is it found in the promoter

Table 3.1. Occurrence of Runs of CG Dinucleotides Found on Searching the
EMBL Nucleotide Sequence Data Library II

	Vertebrate DNA[a]	Nonvertebrate eukaryotic DNA	Random[c]
Bases scanned	216,718	120,003	
Occurrence of			
CG	4,513(20,824)[b]	4,185(34,874)	(40,000)
CGCG	265(1,223)	170(1,417)	(1,600)
CGCGCG	28(129)	9(75)	(64)
CGCGCGCG	5(23)	1(8)	(2.6)
CGCGCGCGCG	2(9)	0	(0.1)

[a] Excluding mitochondrial DNA.
[b] Figures in parentheses indicate frequency per 10^6 bases.
[c] Based on a DNA of 40% G+C content. (The sequence CGCGCGCG was present in a computer-generated random sequence of 247,500 bases at only one-fourth the frequency expected).

region where it might be expected were it to play a role in control of transcription. However the sequence CGTGTGCA occurs twice in the promotor region of a human metallothionein gene and may form Z-DNA, depending on its methylation state (Karin *et al.*, 1984).

It seems highly unlikely that methylation of runs of alternating CG nucleotides could *directly* affect the equilibrium between the B and Z conformers *in vivo*. An indirect action mediated by protein binding to a region of alternating purine and pyrimidine bases is not excluded and would not depend on runs of CG dinucleotides, but on the presence of one CG dinucleotide present in a run of alternating pyrimidine purine residues. Methylation of this cytosine could affect the binding of a protein, which was able to stabilize the region in a Z configuration.

3.5   Other Effects of Cytosine Methylation on DNA Structure

The presence of methylcytosine in DNA affects both the buoyant density and the melting temperature of the DNA (Ehrlich *et al.*, 1975; Szer & Shugar, 1966; Gill *et al.*, 1974; Wagner & Capesius, 1981; Dawid *et al.*, 1970). The increase in stability of double-stranded polynucleotides following substitution of a methyl group for the hydrogen on the 5 position of the cytosine ring is similar to the increased stability found in double-stranded molecules containing thymine in place of uracil (Cassidy *et al.*, 1965). Methylation also increases the stability of triple stranded structures which may form with polypyrimidine:polypurine tracks in the 5' flanking regions of some genes (Lee *et al.*, 1984). This substitution has no effect on the hydrogen-bonding properties of the bases, but may increase the hydrophobic interactions involved in base stacking sufficiently to cause the observed increase in melting temperature. This increase can amount to 6.1° in the *Xanthomonas* phage XP12 in which all the cytosines are replaced by methylcytosine, but is undetectable in other natural polymers where the methylcytosine content is less than 6 mole percent (Schildkraut *et al.*, 1962). Methylcytosine in DNA is considerably more resistant than cytosine to reagents such as bisulfite, which form 5,6 dihydrocytosine adducts (Wang *et al.*, 1980).

The lowering of the buoyant density of DNA substituted with methylcytosine has been elegantly shown by Brown and Dawid (1968) and Dawid *et al.* (1970) in their studies comparing the *Xenopus* ribosomal DNA sequences present in the unmethylated amplified form with the methylated chromosomal material. Here the buoyant density differs by 0.0055 g cm^{-3} and allows clear distinction between the two forms (Figure 3.4). This is only one example of natural DNA's containing methylcytosine having a buoyant density less than expected from a consideration of their G+C content (Ehrlich *et al.*, 1975). The lowering is probably caused by the increase in volume as the methyl group protrudes into the major groove of the DNA duplex and displaces an amount of cesium chloride greater than the increase in weight of the DNA (Kirk, 1967). Theoretical calculations suggest that DNA containing 25% of its cytosines methylated should have a density lower by 0.004 g cm^{-3} than its unmethylated counterpart, and this agrees fairly well with observation. Similar arguments apply to the differ-

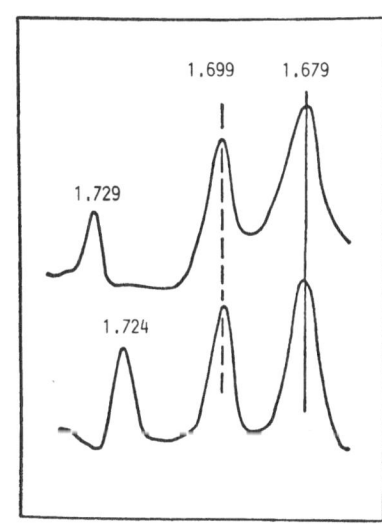

Figure 3.4. CsCl density gradient separations of germinal vesicle ribosomal DNA (top) and somatic cell ribosomal DNA (bottom) of *X. laevis*. Markers of nuclear DNA (density 1.699) and dAT (density 1.679) are present in both runs (from Brown & Dawid, 1968). Copyright 1968 by the AAAS.

ent buoyant densities of DNA containing thymine or uracil. What effect this lowered density might have on any function of DNA is unknown and has not figured in discussions of the role of DNA methylation. There is no effect of methylation on the degree of supercoiling of plasmids (Ciomei *et al.*, 1984).

3.6   Effect of Adenine Methylation on DNA Structure

In contrast to methylation of cytosine, the substitution of 6 methyladenine for adenine in DNA has a major effect on the structure of DNA.

Monomeric N^6-methyladenine can exist in one of two isomeric forms in solution (Figure 3.5; Engel & von Hippel, 1974). Using proton magnetic resonance spectra it was shown that the two isomers interconvert in the millisecond time range but that in solution there is a 20-fold preference for the *cis* over the *trans* rotational isomer. The *cis* isomer is unable to form a Watson-Crick base pair with thymine and the disturbance of the *cis-trans* equilibrium required to form such base pairs in DNA has been calculated to destabilize the double helix by about 1.5 kcal/mol of methylated base pair (Engel & von Hippel, 1978). This destabilization is partly offset by increases in base-stacking interactions caused by the presence of the hydrophobic methyl group, and comparison of melting temperatures shows that the destabilization varies from about 0.35 to about 0.95 kcal/mole methylated base for double helices of different sequences.

Although not actually disruptive of the double helix, the presence of methyladenine does make it more probable that nonpaired regions may exist in methylated DNA, and methylated regions, especially in super-coiled DNA, will have a greater tendency to form hairpin loops. This may play an important part in the mechanism of action of some restriction endonucleases and methylases that recognize palindromic DNA sequences (see Chapter 6). Whereas the unmethylated sequence may be maintained in the regular double helix, sensitive to the endonuclease, the methylated form may adopt an open or cruciform structure that can no longer be cleaved. Methylation of up to 300 adenines in the plasmid pBR322 has only an extremely small effect on helix structure (Cheng *et al.*, 1985).

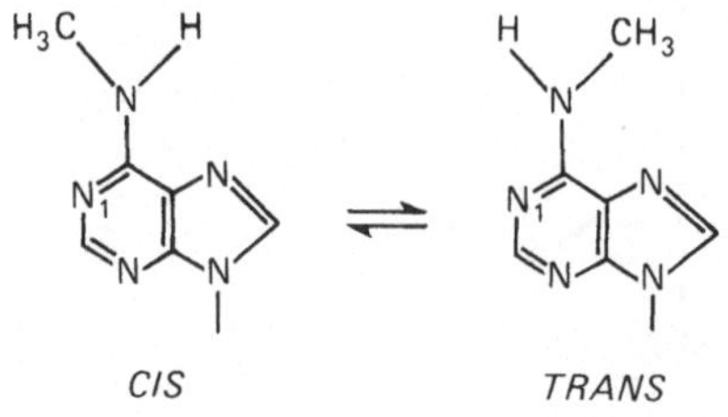

CIS   TRANS

Figure 3.5. The *cis* and *trans* isomers of N^6-methyladenine residues.

Adenine methylation by the *dam* methylase, which methylates adenines in the sequence GATC, is essential for the expression of the phage mu *mom* gene. Upstream from this gene are three GATC sequences that can occur within a complex secondary structure formed by folding DNA sequences back on themselves. Such sequences may interfere with transcription, which only occurs when they are either methylated or deleted (see Section 6.3).

4

S-Adenosyl-L-methionine—Donor of Methyl Groups

4.1 Intracellular Role and Metabolic Cost

The compound S-adenosyl-L-methionine (AdoMet), shown in Figure 4.1, is the methyl group donor in numerous cellular enzymatic transmethylation reactions, including DNA methylation (Cantoni, 1975). In reality AdoMet is probably second only to ATP in the variety of reactions it serves as a cofactor. When it was first discovered, it appeared to contribute only to the modification of small molecules. However it soon became clear that this is a unique modifying agent, created during evolution to shape the structure of macromolecules as well. Proteins, lipids, polysaccharides, and nucleic acids (RNA and DNA) are all subject to methylation by highly specific enzymes after their primary biosynthesis.

The molecular structure of AdoMet is complex, but this is believed to be critical for enzyme-substrate interactions in the catalytic transition state that leads to the large catalytic accelerations that the methyltransferase enzymes achieve. Nevertheless the costs of this elaborate structure are paid for by the individual organisms through the very high energy requirement of AdoMet biogenesis (Cantoni, 1960). This is commented upon by Atkinson (1977), who remarks "in being reduced and activated to form the methyl group of S-adenosylmethionine, a carbon atom from glucose is promoted from an average value of 6.3 [ATP] equivalents (*i.e.,* 38/6) to a cost of 12 equivalents. Active methyl is probably the most expensive metabolic compound or group on a per-carbon basis." This suggests that organisms might experience considerable evolutionary pressure aimed at efficient utilization of AdoMet.

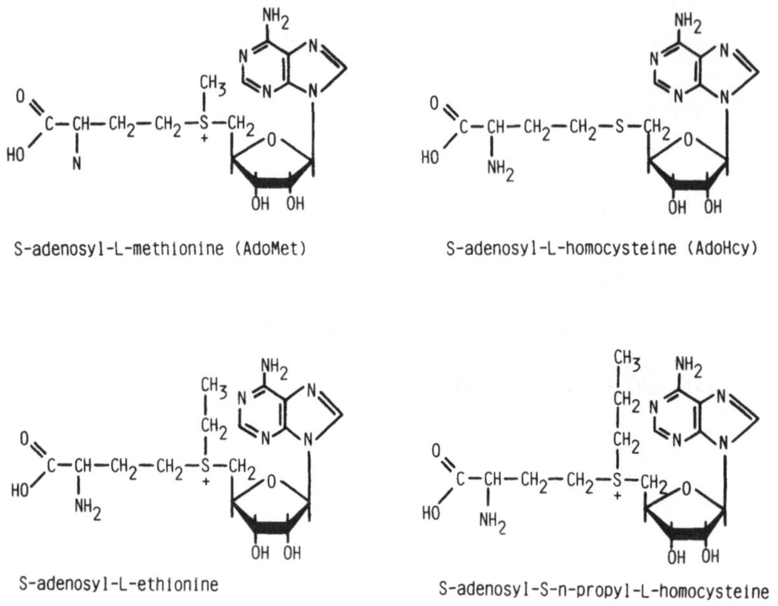

S-adenosyl-L-methionine (AdoMet)

S-adenosyl-L-homocysteine (AdoHcy)

S-adenosyl-L-ethionine

S-adenosyl-S-n-propyl-L-homocysteine

Figure 4.1. Structure of S-adenosyl-L-methionine and related compounds.

4.2 Biosynthesis

S-adenosylmethionine (AdoMet) is formed in a cytosolic enzymic reaction from L-methionine and ATP. The enzyme responsible is S-adenosyl-L-methionine synthetase (ATP:L-methionine S-adenosyltransferase, EC 2.5.1.6), which has been purified from a number of species. With the enzymes from rat liver and yeast, the following stoichiometry was established (see Mudd, 1965).

$$\text{L-methionine} + \text{ATP} \xrightarrow{Mg^{2+}K^+} (-)\text{S-adenosyl-L-methionine} + PP_i + P_i$$

In this reaction a sulfonium compound is formed at the expense of ATP, and maximum activity requires quite high concentrations of both monovalent and divalent ions. Another distinctive feature is that the reaction requires *complete* dephosphorylation of ATP and the nucleophilic transfer of the 5'-deoxyadenosyl moiety of ATP to one of the free pairs of electrons of the thio ether sulfur of L-methionine with the formation of only one of the two diastereoisomeric sulfonium compounds. The source of the inorganic pyrophosphate is the innermost phosphate groups (α and β) of ATP, whereas the terminal and γ-phosphate of ATP gives rise to the inorganic orthophosphate. Mudd (1963), however, has shown that enzymebound tripolyphosphate (PPP_i) is an intermediate in the enzyme reaction, which can be represented as follows (Mudd & Cantoni, 1964).

$$E + ATP \rightleftharpoons E\text{-}ATP$$
$$Met + E\text{-}ATP \rightleftharpoons Met\text{-}E\text{-}ATP$$
$$Met\text{-}E\text{-}ATP \rightleftharpoons AdoMet\text{-}E\text{-}PPP_i$$
$$AdoMet\text{-}E\text{-}PPP \rightleftharpoons AdoMet + E + PP_i + P_i$$

$$\overline{Met + ATP \rightleftharpoons AdoMet + PP_i + P_i}$$

Although this simple formulation of the reactions involved indicates reversibility, only very slow rates of reversal have actually been observed (Mudd & Mann, 1963). Further studies have revealed that both the rat liver and yeast enzymes function only with ATP (Cantoni & Durell, 1957; Mudd & Cantoni, 1958), and it is proposed that there is a *specificity site* on the enzyme that specifically recognizes the adenosyl moiety of the ATP molecule together with an *affinity* site that is complementary to the PPP_i region of the ATP. Because exogenous adenine binds poorly to the enzyme, it has been proposed that in the adenosyltransferase reaction the cleavage of the ribose phosphate bond of ATP and the transfer of the adenosyl group to the sulfur of L-methionine must occur in a concerted process in which the adenosyl group never leaves the enzyme surface.

From inhibitor studies with analogues of L-methionine, Mudd (1965) concluded that the conformation of L-methionine at the active site of adenosyltransferase resembled an extended *trans* configuration. Indeed examination of the effects of L-methionine and certain nonsubstrate analogues on the reaction velocity of the rat liver enzyme revealed a concentration-dependent biphasic activation-inhibition phenomenon. Low concentrations of analogues such as cycloleucine or L-2-amino-4-hexynoic acid activated the enzyme, whereas higher concentrations were inhibitory (competitive with L-methionine).

In general, the adenosyltransferase enzymes studied so far from rat liver, rat kidney, human erythrocytes, yeast, and *E. coli* differ considerably in their kinetic properties and molecular weights (Hoffman & Kunz, 1980; Hoffman & Sullivan, 1981; Abe *et al.*, 1982; Okada *et al.*, 1981; Oden & Clarke, 1983; Chiang & Cantoni, 1977; Markham *et al.*, 1980). Whereas the rat liver activity appears as two (or even three) isoenzymes, the others seem to exist as single molecular species. One of the rat liver isoenzymes (K_m methionine 200 μM) is activated by its product, AdoMet. On the other hand, the human red cell enzyme, the other rat liver isoenzyme(s), and the *E. coli* enzyme show varying degrees of product inhibition (see Oden & Clarke, 1983). For example, the red cell enzyme (K_m methionine 2.2 μM) is strongly inhibited by AdoMet (K_i 2.0–2.9 μM), the endogenous AdoMet concentration in erythrocytes being 3.5 μM.

4.3   Adenosyltransferase and Regulation of Intracellular Concentration

The observations regarding the regulation of adenosyltransferase activity mentioned in Section 4.2 raise the question of the potential role of this enzyme in regulating cellular AdoMet concentration. A close relationship between intracellular levels of L-methionine and of AdoMet was pointed out by Schlenk and his collaborators (Schlenk *et al.*, 1965; Svihla & Schlenk, 1960) who demonstrated accumulation of AdoMet in yeast cells incubated in media rich in methionine. In rats, the administration of methionine also led to increases in AdoMet in all tissues (Baldessarini & Kopin, 1966). Such observations suggested that the AdoMet levels are controlled by rates of synthesis rather than rates of utilization and that rates of synthesis are limited by tissue levels of L-methionine (Lombardini *et al.*, 1971). In rat liver and yeast the intracellular level of L-methionine are well below the K_m's reported for the adenosyltransferases from these sources. When the methionine analogue inhibitors mentioned previously (*e.g.*, cycloleucine, L-2-amino-4-hexynoic acid) are administered to rats or mice *in vivo,* the level of methionine (normally 50–180 nmoles/g wet tissue weight) rises by two- to fourfold (Chou *et al.*, 1977). Concomitantly there is a decline in tissue levels of AdoMet (see Table 4.1 for normal values) in spleen, kidney, pancreas, brain, and adrenals but not in liver (Chou *et al.*, 1977). Liver, on the other hand, is the richest source of adenosyltransferase, having 20 times the level of other tissues (Chou *et al.*, 1977), and the presence of large quantities of enzyme could result in the maintenance of considerable rates of AdoMet synthesis in rat liver.

Table 4.1 Estimate of S-Adenosylmethionine Content in Various Rat Tissues[a]

Tissue	S-Adenosylmethionine (nmol/g wet tissue weight)
Brain	23.5 ± 1.3
Heart	34.5 ± 5.0
Kidney	52.3 ± 3.4
Lung	36.0 ± 4.7
Small intestine	28.2 ± 4.3
Spleen	46.4 ± 3.8
Prostate	57.4 ± 4.6
Liver	66.9 ± 7.1

[a] Data are from Hibasami *et al.* (1980). (Results are shown as mean ± SD for at least 4 determinations from different animals.)

In human red blood cells, the low K_m exhibited by the adenosyltransferase for methionine (2.2 μM) and the larger physiological intracellular concentration of methionine (12–24 μM) indicate that in these cells, unlike the situation mentioned above, the enzyme is likely to be saturated with this substrate (Oden & Clarke, 1983). Thus increases in methionine concentration would not lead to greater enzyme activity. This seems to be the case, as red cell levels of AdoMet are not markedly altered by oral feeding or perfusion of large doses of methionine. On the other hand, it has been proposed by Oden and Clarke (1983) that the specific properties of the red blood cell adenosyltransferase are related to a specific role in erythrocyte methyl group metabolism, especially with regard to protein carboxyl methyl-transfer reactions.

In general, therefore, it appears that the enzyme adenosyltransferase plays a key role not only in the formation of AdoMet but also in the regulation of intracellular levels of AdoMet. While the levels of AdoMet in various rat tissues are presented in Table 4.1 and appear to vary from tissue to tissue, there are also data to indicate that variations occur with growth rate. For instance the intracellular content and turnover of AdoMet has been measured in cultured human lymphocytes (German *et al.,* 1983). In exponentially growing W1-L2 lymphoblasts, the AdoMet level is found to be 59 nmol/ml cell volume, whereas in unstimulated peripheral blood mononuclear cells, the level is only 3.3 to 8.1 nmol/ml cell volume. However, after stimulation with phytohemagglutinin for 24 to 48 hours the level of AdoMet in these cells rises to 25 to 33 nmol/ml cell volume. Additional data (German *et al.,* 1983) showed that AdoMet biosynthesis accounts for 20 to 23% of the L-methionine utilization in the W1-L2 cells. Moreover, in these cells and in the lectin-stimulated mononuclear cells, approximately 82 to 88% of that AdoMet is used in methyltransferase reactions.

4.4 Stability and Its Significance

As already mentioned, the product of the action of the enzyme ATP:L-methionine S-adenosyltransferase is (-)S-adenosylmethionine (Section 4.2). This enzymically active stereoisomer, 5'-[[(3S)-3-amino-3-carboxy-propyl]methyl-(S)-sulfonic]-5'deoxyadenosine, is also referred to as (-)-AdoMet or (S,S)-AdoMet. This molecule can hydrolyze even at neutral pH by a mechanistically complex route to 5'-deoxy-5-(methylthio)-adenosine (MTA) and homoserine (Parks & Schlenk, 1958; Mather, 1958; Mudd, 1959; Zappia *et al.,* 1977b, 1979). A possible mechanism (Wu *et al.,* 1983) is shown in Figure 4.2, which depends on the presence of the carboxyl group. The reaction is proposed to proceed by intramolecular cyclization followed by hydrolysis of the lactone. In addition, as shown in

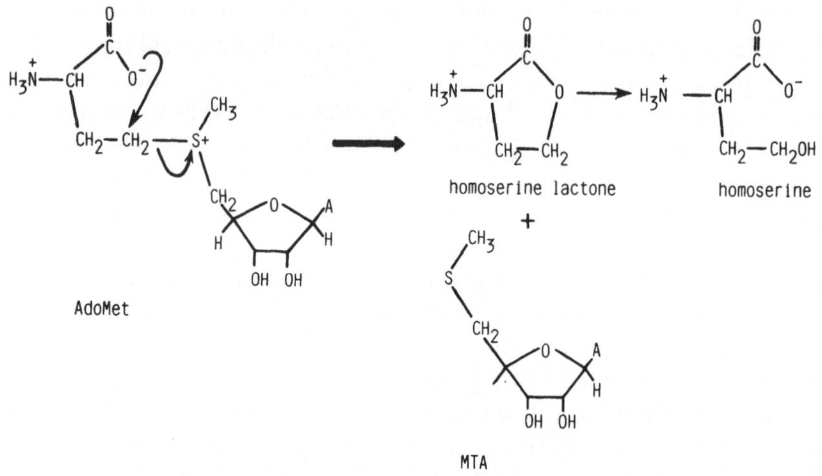

Figure 4.2. Hydrolysis of S-adenosyl-L-methionine.

Figure 4.3, the S chirality at the sulfonium pole can be lost through spontaneous racemization to yield the enzymically inactive stereoisomer (R,S)-AdoMet (Wu *et al.*, 1983). Indeed, it is reported that commercial samples of AdoMet from biological sources can contain as much as 20% of this inactive stereoisomer (Stolowitz & Minch, 1981).

At 37° at pH 7.5, the *racemization* rate constant for AdoMet is 8×10^{-6} s^{-1} (Wu *et al.*, 1983). On the other hand, the rate constant for the *hydrolysis* of AdoMet to MTA and homoserine at 37° and pH 7.5 is 6×10^{-6} s^{-1}. Thus, under physiological conditions (Wu *et al.*, 1983), (S,S)-AdoMet decays to an equal mixture with its inactive disastereomer with a half-life of approximately 12 hours. Under the same conditions, (S,S)-AdoMet is lost through irreversible MTA formation with a half-life of 32 hours.

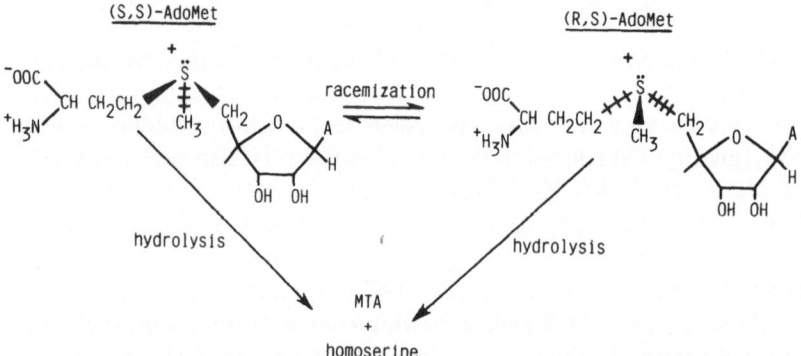

Figure 4.3. Racemization and hydrolysis of S-adenosyl-L-methionine.

With these chemical data in mind, Wu *et al.* (1983) reconsidered the 12 ATP synthetic cost of (S,S)-AdoMet formation. They calculated that within a period of say 4 hours, roughly 1 ATP equivalent of (S,S)-AdoMet will be lost through hydrolysis to MTA. For reasons of cellular economy, they suggested that all cellular AdoMet must either have a rapid metabolic turnover, or some binding moiety capable of chemical stabilization of AdoMet must be present within cells. It turns out that in human liver the half-life of AdoMet methyl groups is 3.5 to 9 minutes (Mudd *et al.*, 1980), and in rat liver the half-life is 10 minutes (Baldessarini, 1975). Moreover, the existence of an AdoMet binding protein has been reported (Smith, 1976).

Wu *et al.* (1983) also considered the racemization and indicated that 1 ATP equivalent of (S,S)-AdoMet would be converted to (R,S)-AdoMet in 3 hours. The loss in this case is not permanent. Whereas (S,S)-AdoMet can be consumed enzymically by the various cellular methyltransferases (*e.g.*, DNA methylase), the (R,S)-AdoMet will revert to (S,S)-AdoMet albeit with a half-life of 12 hours. Indeed (R,S)-AdoMet may be considered as a slow-release storage form of AdoMet. A problem however would be that a sudden demand for (S,S)-AdoMet would put constraints on cellular transmethylation rates. Of course, again a cellular binding protein would be useful if it could protect AdoMet against a loss of chirality.

In summary, then, the instability of (S,S)-AdoMet with respect to loss of chirality and spontaneous hydrolysis places considerable constraints on any organism that employs it. This could be circumvented by high turnover rates or stabilization through macromolecular binding in order to avoid gross waste of metabolic energy.

While the above-mentioned chemical changes can occur to AdoMet under near physiological conditions, treatment at pH 2 will result in the further breakdown of MTA to adenine and methylthioribose. Alternatively, treatment with mild alkali, *even at* 0°C, will yield adenine and S-ribosylmethionine. If hot alkali is used, then adenine, ribose, and methionine will result (see Michelson, 1963).

4.5 Analogues

Attempts to prepare analogues and derivatives of AdoMet by biosynthesis using the adenosyltransferase enzyme described in Section 4.2 have been made difficult by virtue of the high specificity of the enzyme (Lombardini *et al.*, 1970; Coulter *et al.*, 1974). Mudd and Cantoni (1957) were able to show that selenomethionine can react with ATP in a fashion analogous to methionine in AdoMet biosynthesis, but the selenium compound is unstable.

More progress, however, has been made with intact cells. For example, with intact yeast cells, L-ethionine was found to be as effective a precursor as L-methionine (Schlenk *et al.*, 1959, 1965). Possibly, intact cells can remove the product from its site of synthesis, or perhaps the enzyme in the cell has a less-stringent specificity. Whatever the reason, this approach with intact cells (Schlenk, 1977) has yielded analogues such as S-adenosyl-L-ethionine (using L-ethionine) and S-adenosyl-(S-*n*-propyl)-L-homocysteine, the S-*n*-propyl analogue (using S-*n*-propyl-L-homocysteine) (see Figure 4.1).

When these analogues are tested as substrates for methyltransfer enzymes, the rate of enzymatic alkyl transfer declines dramatically with increase in size from the methyl to the ethyl and *n*-propyl group (Schlenk, 1977).

4.6   Catabolism

4.6.1   S-Adenosylhomocysteine Formation

All enzymic transmethylation reactions in which methyl groups are transferred from AdoMet to an acceptor (*e.g.*, cytosine residues of DNA) yield S-adenosyl-L-L-homocysteine (AdoHcy).

$$\text{AdoMet} + \text{A (methyl group acceptor)} \xrightarrow{\text{methyltransferase}} \text{AdoHcy} + \text{A-CH}_3$$

AdoHcy was first characterized in 1954 (Cantoni & Scarano, 1954), and a key feature of this particular product is that it acts as an extremely powerful inhibitor of methyltransferase catalyzed reactions (Zappia *et al.*, 1969; Deguchi & Barchas, 1971; Hurwitz *et al.*, 1964; Kerr, 1972; Pegg, 1971; Hildesheim *et al.*, 1973) (see Section 5.4.1).

4.6.2   Hydrolysis of S-Adenosylhomocysteine

In eukaryotic cells at least, AdoHcy can be cleaved by the enzyme S-adenosyl-L-homocysteine hydrolase (EC 3.3.13) (De la Haba & Cantoni, 1959; Duerre & Walker, 1977).

$$\text{AdoHcy} + \text{H}_2\text{O} \xrightarrow{\text{hydrolase}} \text{L-homocysteine} + \text{adenine}$$

A distinctive feature of this particular enzyme reaction is that equilibrium lies far in the direction of condensation, however AdoHcy will undergo hydrolysis if the products of the hydrolytic reaction, homocysteine and adenine, are removed. Conveniently, these two metabolites are continuously utilized by other metabolic pathways within the intact cell, and thus *in vivo,* the enzymic reaction is therefore toward the hydrolysis of AdoHcy (Ueland, 1982). Adenosine is either phosphorylated to AMP or deaminated to inosine, while Hcy is metabolized to cystathionine or re-methylated to methionine. The latter reaction is catalyzed by two enzymes, one of which is widely distributed in mammalian tissues and requires 5'-methyltetrahydrofolate as methyl donor (Ueland, 1982; Mudd & Levy, 1983; Ueland *et al.,* 1984). The cellular metabolism of adenine or homocysteine can, however, be affected, leading to deleterious increases in the level of AdoHcy (Ueland, 1982). For example, in cultured S49 mouse lymphoma cells, treatment with erythro-9-(2-hydroxy-3-nonyl)-adenine, a specific inhibitor of adenosine deaminase, leads to increased levels of adenosine, which in turn leads to levels of AdoHcy that are toxic and that severely inhibit DNA methylation (Kredich & Martin, 1977) (see Figure 5.14). Excess adenosine ($>15 \mu M$) in the medium will also elicit the same effect (Kredich & Martin, 1977) as does the addition of exogenous L-homocysteine (Kredich & Martin, 1977; Hoffman *et al.,* 1980).

The AdoHcy hydrolase itself is readily detected in a number of mammalian organs. For example in the rat, the highest levels of activity are found in the liver, kidney, and pancreas, lower in spleen and testes, and very low in brain and heart (Duerre & Walker, 1977). In mice, the enzyme can be nearly completely inactivated by injection of the drug combination 9-β-D-arabinofuranosyl adenine plus 2-deoxycoformycin. Such an inhibitor results in a massive accumulation of AdoHcy in all tissues (Helland & Ueland, 1983).

While AdoHcy hydrolase with similar properties to the mammalian enzymes has been detected in yeast, a somewhat different situation exists in prokaryotes. In bacteria, AdoHcy is metabolized *irreversibly* by a two-step reaction. The first is catalyzed by AdoHcy nucleotidase, resulting in the formation of adenine and S-ribosylhomocysteine. The latter compound is then further split by an S-ribosylhomocysteine cleavage enzyme to yield free homocysteine (Duerre & Walker, 1977) (see also Figure 4.5).

4.6.3 Decarboxylation of S-Adenosylmethionine

While AdoMet serves as a substrate for methyltransferase reactions with the resulting production of AdoHcy, it should be pointed out that AdoMet can also be decarboxylated enzymatically. The enzyme responsible, S-adenosylmethionine decarboxylase, has been reported to occur in a wide variety of cell types (Zappia *et al.,* 1977a).

$$(-)\text{AdoMet} \xrightarrow{\text{AdoMet decarboxylase}} \text{decarboxylated AdoMet} + CO_2$$

Whereas the decarboxylase from *E. coli* requires Mg^{2+} and pyruvate as prosthetic group (Wickner *et al.,* 1970), the enzymes from rat liver, rat

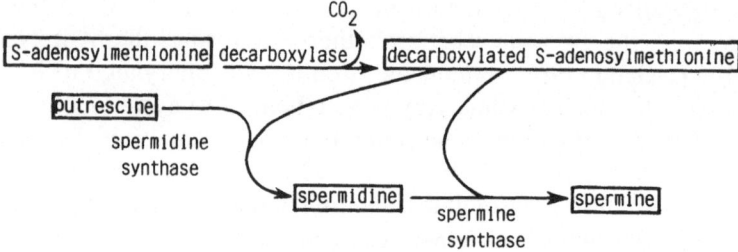

Figure 4.4. S-adenosyl-L-methionine and polyamine biosynthesis.

prostate, and yeast are activated by putrescine but have no requirement for Mg^{2+} (Williams-Ashman *et al.*, 1977; Pegg, 1974; Williams-Ashman & Pegg, 1980). Only (S,S)AdoMet or (-)AdoMet stereoisomer is decarboxylated, and AdoHcy cannot serve as a substrate. On the other hand, unlike the methyltransferases, the AdoMet decarboxylase does not appear to be inhibited by AdoHcy. A potent inhibitor, however, of mammalian decarboxylases is the antileukemia agent, methyl glyoxal bis(guanylhydrazone) (MGBG). Its effect on the *E. coli* enzyme is not so strong.

The intracellular concentration of decarboxylated AdoMet in mammalian cells has been estimated as only 0.9 to 2.5% of the level of AdoMet (see Hibasami *et al.*, 1980). Nevertheless, it is this decarboxylated product that is the key intermediate in the synthesis of polyamines in prokaryotes and eukaryotes, being the donor of the vital aminopropyl groups transferred by spermidine and spermine synthetases, as shown in Figure 4.4 (Tabor & Tabor, 1972).

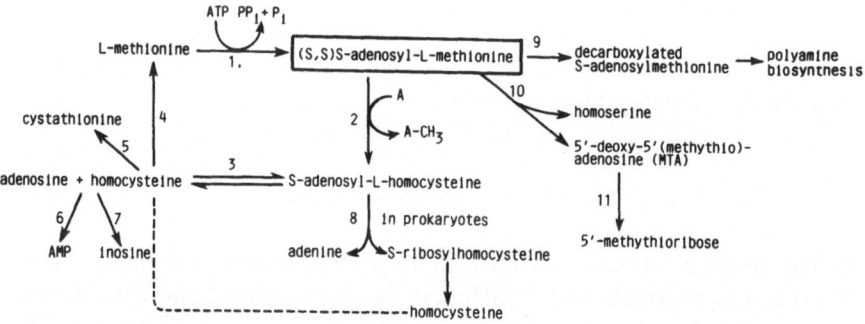

Figure 4.5. Metabolism of S-adenosyl-L-methionine. (1) ATP:L-methionine S-adenosyltransferase; (2) methyltransferases (*e.g.*, DNA methylase); (3) S-adenosyl-L-homocysteine hydrolase (eukaryotes); (4) 5-methyltetrahydrofolate:L-homocysteine methyltransferase; (5) cystathionine-β-synthetase; (6) adenosine kinase; (7) adenosine deaminase; (8) S-adenosyl-L-homocysteine nucleotidase (prokaryotes); (9) S-adenosyl-L-methionine decarboxylase; (10) S-adenosyl-L-methionine lyase; (11) methylthioadenosine nucleotidase.

4.6.4   S-Adenosyl-L-Methionine Lyase

Although AdoMet can hydrolyze spontaneously (see Section 4.4), bacteria and yeast at least have been shown to contain a lyase that will catalyze the conversion of AdoMet to MTA plus homoserine lactose. The MTA is then broken down to methylthioribose and adenine by MTA nucleotidase (see Figure 4.5).

4.6.5   Summary

Figure 4.5 presents a summary of the above-mentioned AdoMet catabolic pathways in diagrammatic form.

5

The Methylation Machinery: DNA Methyltransferases

DNA methyltransferases (DNA methylases) transfer methyl groups from S-adenosylmethionine (AdoMet) to certain adenine or cytosine bases in polymeric DNA (see Chapter 2). In prokaryotes, most methylases are associated with endonucleases as part of restriction-modification systems (see Chapter 6) though there also exist the *dam* and *dcm* methylases, which do not have associated endonucleases. In eukaryotes, the only methyltransferases characterized are those from higher eukaryotes and from *Chlamydomonas*, which methylate cytosine bases in DNA.

5.1. Prokaryote Methylases

Bacterial restriction endonucleases fall into three classes. In type II, the simplest, the endonuclease and its corresponding methylase both contain a single type of polypeptide, but each acts independently of the other. They require Mg^{2+} for endonuclease action and AdoMet for modification. At least in *E. coli*, the type II restriction-modification systems are encoded on plasmids, and the genes for the nuclease and methylase are adjacent to one another. The type III restriction endonucleases have two nonidentical subunits and require Mg^{2+} and ATP and are stimulated by AdoMet. However, the smaller subunits can act alone as modification methylases similar to the type II methylases, except for a requirement for Mg^{2+}.

Class I restriction enzymes are the most complicated and contain three nonidentical subunits coded by three contiguous genes on the *E. coli* chromosome. For endonuclease and methylase activity they require Mg^{2+},

ATP, and AdoMet, but only two of the subunits are required for methyl-ase activity.

In addition to the methylases associated with restriction enzymes there are, in *E. coli,* the *dam* and *dcm* methylases, at least the former of which is a single polypeptide chain resembling the type II methylases. There are also methylases separately induced by viruses and these are not found in association with restriction endonucleases (Gunthert & Trautner, 1984; Shlagman & Hattman, 1983).

In the case of the type II enzymes, both the nuclease and the methylase independently recognize the same short oligonucleotide sequence as a preliminary to either cleaving or methylating the DNA. Because of this, it might be expected that the proteins would show some structural similarity at least over the region involved in DNA recognition. Furthermore, the finding that the EcoRI endonuclease and methylase genes are adjacent to each other [as are the genes for all Class II pairs of cognate endonuclease and methylase so far investigated except the MspI pair (Walder *et al.,* 1983)] suggests that they may have arisen as a result of a gene duplication event, but comparison of the nucleotide sequence of the two genes shows insignificant homology. The same is true for the amino acid sequence inferred from the nucleotide sequence. The lack of homology also extends to higher orders of structure as judged by circular dichroism and predic-tion of secondary structure by probabilistic methods (Newman *et al.,* 1981; Greene *et al.,* 1981). One must conclude then that quite different protein structures can recognize identical oligonucleotide sequences and also evolve to produce the two enzymes coded for by adjacent plasmid genes involved in the coupled reactions of restriction and modification. This conclusion that there is no evolutionary relationship between the proteins is hard to swallow, and it is still possible that the two genes did arise from a common ancestor but have diverged so much that we can no longer detect any sequence relationship between them. Work with cog-nate endonucleases and methylases suggests, however, that the two en-zymes do not share common contacts with their DNA substrate (Yoo *et al.,* 1982; Modrich & Roberts, 1982).

5.1.1   Type II Methyltransferases[1]

A number of type II methylases that transfer methyl groups to either N6 of adenine or to the 5 position of cytosine have now been purified (Table 5.1). The reactions catalyzed are described in Chapter 2.

All the methylases except that from *Bacillus stearothermophilus,* which has a subunit molecular weight of 105,000 and functions as a membrane-bound tetramer, are in the size range 30,000 to 40,000 and exist as mono-

---

[1] Modrich & Roberts, 1982.

Table 5.1. Bacterial Methylases Purified

Methylase	Source	Recognition sequence	Reference
*Eco*RI	*E. coli*	GAȦTTC	Rubin & Modrich, 1977
*Hpa*I	*H. parainfluenzae*	GTTAȦC	Yoo *et al.*, 1982
*Hpa*II	*H. parainfluenzae*	CĊGG	Yoo & Argarwal, 1980
*Bsu*RI a and b	*B. subtilis*	GGĊC	Gunthert *et al.*, 1981a
Bst 1503 I	*B. stearothermophilus*	GGȦTCC	Levy & Walker, 1981
*Bam*HI	*B. amyloliquefaciens*	GGATĊC	Nardone *et al.*, 1984
dam	*E. coli*	GȦTC	Geier & Modrich, 1979
*Eco*B	*E. coli*		Horiuchi & Zinder, 1972
*Eco*K	*E. coli*		Yuan *et al.*, 1980
*Eco*P15	*E. coli*		Hadi *et al.*, 1983 Reiser & Yuan, 1977
*Eco*PI	*E. coli*		Hadi *et al.*, 1983
Hinf III	*H. influenzae*		Kauc & Piekarowicz, 1978

mers in solution. This is in contrast to the corresponding endonucleases, which usually interact with DNA as dimers, a fact considered vital to their action on a symmetrical tetra-, penta-, or hexanucleotide sequence. This difference implies that the methylase and endonuclease interact with DNA in different ways, which is a further argument against a common evolutionary origin. The cloning of methylase genes (Table 5.2) is already enabling information to be gleaned about the control of their transcription as well as about the structure of the encoded proteins. It appears that the EcoRI endonuclease and methylase genes, although contiguous, are transcribed from different promoters (O'Connor & Humphreys, 1982). More-

Table 5.2. Bacterial Methylase Genes Cloned

*Eco*RI	Newman *et al.*, 1981; Greene *et al.*, 1981; O'Connor & Humphreys, 1982
*Eco*RV	Bougueleret *et al.*, 1984
*Msp*I	Walder *et al.*, 1983
*Bsp*I	Szomolanyi *et al.*, 1980
*Hha*II	Mann *et al.*, 1978
*Eco*RII	Kosykh *et al.*, 1980
*Pst*I	Walder *et al.*, 1981
*Bsu*RI	Gunthert & Trautner, 1984
dam	Brooks *et al.*, 1983, Herman & Modrich, 1982
*Eco*K	Sain & Murray, 1980
*Eco*PI	Iida *et al.*, 1983

over, the availability of the pure proteins will allow their interaction with DNA to be studied in greater detail.

5.1.1a Site of Action of Methylase

The location of methylcytosines added *in vitro* by purified type II methylases has been investigated by direct sequencing using the Maxam-Gilbert method (Walder *et al.*, 1983; Gunthert *et al.*, 1981b) or by sensitivity to a series of restriction endonucleases whose specificity is known (Quint & Cedar, 1981). Simple polynucleotides can be used as substrates in certain cases. For example, the alternating copolymer poly d(AT) is a substrate for EcoRI methylase showing the reduced specificity of this enzyme under modified incubation conditions (Woodbury *et al.*, 1980). The site of action of the *Hpa*I methylase was found by methylating the labeled octanucleotide [5'-^{32}P]d(G-G-T-T-A-A-C-C). After treatment with *Hpa*I methylase and [^{3}H]AdoMet, the octanucleotide was partially cleaved by snake venom phosphodiesterase to generate all possible fragments. These were separated chromatographically, and the longest double-labeled fragment found was the hexanucleotide. This clearly shows the 3' adenosine to be the methyl acceptor (Yoo *et al.*, 1982).

Earlier approaches involved digestion of DNA methylated *in vitro* with a series of endonucleases to give a mixture of dinucleoside monophosphates, which were separated by chromatography or electrophoresis. Following methylation by *Eco*RII methylase and [^{3}H]AdoMet, three radioactive dinucleotide spots were located, which were further treated with spleen phosphodiesterase (to give Xp + Y) or venom phosphodiesterase (to give X + pY). This enabled the labeled dinucleotides to be identified as Cpm^{5}C, m^{5}CpA, and m^{5}CpT present in the ratio 2:1:1. Thus, the methylated sequence is CmCA/T. Similar treatment of radiolabeled trinucleoside diphosphates confirms the full methylated sequence as CmCA/TGG (Boyer *et al.*, 1973; Roy & Smith, 1973).

5.1.1b Purification and Assay of Type II Methylases

No unusual procedures have been used in the purification of type II DNA methylases. A combination of ion exchange chromatography, gel filtration, and affinity chromatography on DNA or heparin agarose yields enzyme preparations with specific activities of about 10^6 units per mg protein (a unit transfers one pmole methyl group to DNA per hour). The assay involves incubating [^{3}H]AdoMet with DNA and any necessary cofactors and then removing excess unincorporated [^{3}H]AdoMet (see Section 5.2.2). Alternative assays involve following the appearance of DNA resistant to the cognate restriction endonuclease. Thus Bst 1503I endonu-

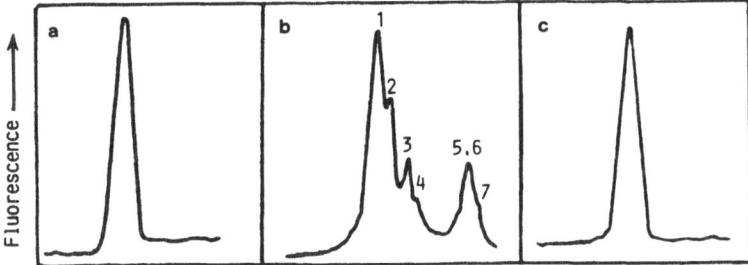

Figure 5.1. Methylation of phage TP-1C DNA by Bst 1503I methylase. Agarose gel electrophoresis separations are shown of (a) unmethylated DNA; (b) unmethylated DNA incubated with the cognate endonuclease; (c) DNA incubated for 90 minutes with the methylase followed by the cognate endonuclease. (Reprinted with permission from Levy & Walker, *Biochemistry 20:*1120–1127. Copyright 1981 American Chemical Society.)

cleave cuts phage TP-IC DNA into seven fragments, the three smallest (5, 6, and 7) running together on electrophoresis, as shown in Figure 5.1. Prior treatment with Bst 1503 I methylase reduces the amount of the small fragments produced. In this assay, one unit of methylase activity is defined as the amount of enzyme required to produce a 1% decrease in the amount of fragments 5, 6, and 7 in 1 minute (Levy & Walker, 1981).

5.1.1c Properties and Mechanism of Action of Type II Methylases

Except for the methylase from *B. stearothermophilus,* which exhibits maximal activity at about 60° and pH 9 (it also differs in size and location, as discussed above), most type II methylases have an optimum temperature around 40° and an optimum pH between 7 and 8. They are inhibited by NaCl concentrations above 100 mM and are somewhat stimulated by EDTA. Sulfhydryl groups are essential for their action, which is blocked by N-ethylmaleimide.

Type II methylases show much greater activity with a double-stranded DNA substrate than with single-stranded DNA. The rate of methylation of unmethylated DNA is constant until the DNA is saturated, indicating that one methyl group is transferred per reaction step by a monomeric enzyme and that each methylation occurs independently of the next. When this is measured by a nuclease sensitivity assay, resistance increases in proportion to methyl groups added at least until 40% saturation, as shown in Figure 5.2. This confirms the separate addition of methyl groups to both sides of the symmetrical site. As the rate of conversion to resistant DNA slows as saturation approaches, this implies that addition to hemimethylated sites is less favorable than to completely unmethylated

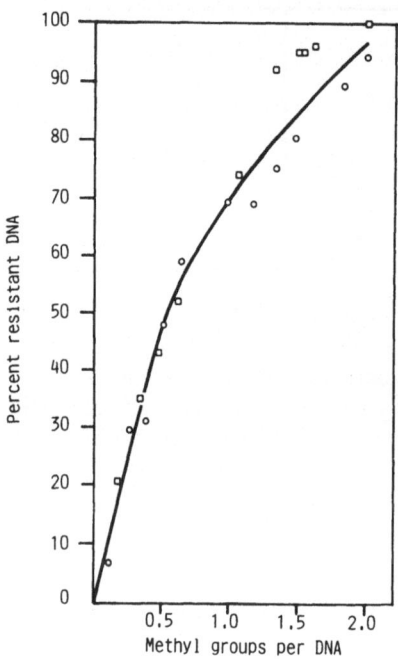

Figure 5.2. *Eco*RI methylase transfers one methyl group at a time. pVH51 DNA was methylated with *Eco*RI methylase. At various times, samples were removed to assess the extent of methylation and the proportion of molecules resistant to *Eco*RI endonuclease. The initial straight line has a slope of 95% resistance per methyl group per DNA molecule, demonstrating that the enzyme transfers methyl groups one at a time and leaves the DNA after each catalytic event. Results of two similar experiments are shown. (From Rubin & Modrich, 1977.)

sites. This is not a universal finding, though the rates are never very different (Rubin & Modrich, 1977; Yoo *et al.*, 1982; Gunthert & Trantner 1984; Gunthert *et al.*, 1981b; Nardone *et al.*, 1984).

Using equilibrium dialysis and competition with analogues, Gunthert *et al.* (1981b) concluded that the BsuRI methylase interacts first with Ado-Met and then nonspecifically with DNA. When a recognition site is encountered, transfer of a methyl group occurs at 37°C and the complex dissociates.

1. E + AdoMet → E-AdoMet
2. E-Ado-Met + DNA → DNA ~ E-AdoMet
3. DNA ~ E-AdoMet → DNA-(GGCC)-E-AdoMet
4. DNA-(GGCC)-E-AdoMet → DNA-(GGmCC)-E-AdoHcy
5. DNA-(GGmCC)-E-AdoHcy → DNA-(GGmCC) + E + AdoHcy

The K_m values for AdoMet are in the range 0.1 μM to 0.7 μM and for DNA between 1 and 90 nM in terms of acceptor sites.

The turnover numbers for type II methylases are low. HpaI methylase will only transfer one methyl group per minute per monomer at 37°C. The *Eco dam* methylase will transfer 19 methyl groups per minute per monomer at 37°C. Nonetheless, there is sufficient enzyme in a cell to fully methylate newly synthesized DNA very shortly after replication (see Section 2.2), and *in vitro* purified enzyme preparations can be used to saturate unmethylated DNA within a few minutes.

5.1.2 The *Eco dam* Methylase

This enzyme has been purified to a specific activity of about 2 million units per mg protein by chromatography on alumina Cγ gel, phenylalanine Sepharose, phosphocellulose, Blue dextran Sepharose, and tRNA agarose (Geier & Modrich, 1979; Herman & Modrich, 1982). This was facilitated by starting with an extract of a clone that overproduces the enzyme 10- to 20-fold. The enzyme shows greater activity at about pH 8 and at NaCl concentrations below 100 mM. It is inhibited by N-ethylmaleimide.

The *Eco dam* methylase is a single polypeptide chain of molecular weight 31,000 comprising 278 amino acids. It functions as a monomer and has a K_m for AdoMet of 12.2 μM and for d(GATC) sites of 3.6 nM. Like the type II methylases, it transfers one methyl group at a time (to the adenine on the sequence GATC) prior to dissociating from the DNA, though it does show a slight preference for hemimethylated sites. It acts only slowly on denatured DNA and this activity has been attributed to the ability of the *Eco dam* methylase to pair single-stranded DNA sequences even within very short palindromic stretches (Buryanov *et al.*, 1984). Thus the enzyme shows virtually no activity with duplex B (below), which is prevented from forming a double-stranded recognition site but is active with duplex A and with either of the separated constituent oligonucleotides that can self anneal over the recognition site.

Duplex A: C A G T T T A G $\boxed{\text{G A T C}}$ C A T T T C A C

            G T C A A A T C $\boxed{\text{C T A G}}$ G T A A A G T G

Duplex B: C A G T T T A G $\boxed{\text{G A T C}}$ C A T T T C A C

            G T C A A A T C C T   G G T A A A G T G

All members of the family Enterobacteriaceae and Haemophilus tested by Brooks *et al.* (1983), together with several other species, have a functional *dam* methylase. Methylases of similar specificity are induced by phages T2 and T4 (van Ormondt *et al.*, 1975; Schlagman & Hattman, 1983). This conservation is quite different from the situation with the type II restriction-modification enzymes and reinforces the suggestion that *dam* methylation plays a specific role in bacterial metabolism (see Section 6.2).

5.1.3 Type I Methyltransferases[2]

The type I restriction-modification systems found in *E. coli* B, *E. coli* K12, and in different *Salmonella* species all share homologies and are

---

[2] Yuan, 1981.

```
Eco K        5'-   A A C N N N N N N G T G C   -3'
             3'-   T T G N N N N N N C A C G   -5'

Eco B        5'-  T G A N N N N N N N N N T G C T  -3'
             3'-  A C T N N N N N N N N N A C G A  -5'
```

Figure 5.3. The recognition nucleotide sequences for the *Eco*K and *Eco*B methylases.

coded for by three contiguous genes known as *hsd*R, *hsd*M, and *hsd*S (standing for *host specificity for DNA* Restriction, Modification, and Specificity). While all three gene products are required for restriction, only *hsd*M and *hsd*S products are required for modification. The *hsd*S gene product is required to recognize the specific sequence in DNA that is modified by the *hsd*M gene product. The *hsd*S and M gene products from *E. coli* B have molecular weights of 60,000 and 55,000 and associate with the largest subunit (molecular weight 135,000) to form the restriction endonuclease (Lautenberger & Linn, 1972; Suri *et al.*, 1984). The different subunits associate in a 2 : 2 : 1 ratio in *E. coli* K but in a 1 : 2 : 1 and a 1 : 1 : 1 ratio in *E. coli* B. In the latter, a separate methylase has been detected containing one each of the two smaller subunits, *i.e.*, 1 : 1 : 0 (Eskin & Linn, 1972; Lautenberger & Linn, 1972). The purified enzymes are very sensitive to -SH oxidation and are maintained in 0.5 mM dithiothreitol. The specific activity is only about 1000 units per mg protein but activity is very dependent on the DNA substrate.

Both the *Eco*B and *Eco*K methylases methylate adenine residues in the complex sequence shown in Figure 5.3 (Kan *et al.*, 1979), *i.e.*, there are seven specified bases in two groups separated by a turn of the double helix so the two modified adenines are adjacent. Restriction takes place several thousand bases away from the recognition and methylation sequence.

The complete restriction-modification enzyme binds five or more molecules of AdoMet and is converted to an activated form that can bind to any DNA molecule. Probably by sliding along the DNA, the enzyme-AdoMet complex now locates itself firmly on the recognition sequence, as shown in Figure 5.4. The recognition sequence can exist in one of three forms and subsequent activity depends on which is present:

1. Site fully methylated on both strands—enzyme is released in an ATP-dependent reaction.
2. Hemimethylated site—the unmethylated adenine is methylated in an ATP-stimulated reaction and then the enzyme is released.
3. Unmethylated site—In the absence of ATP, a slow methylation reaction can be measured, but in the presence of ATP, a conformation change occurs to the enzyme, AdoMet is released, and a translocation

event occurs, which results in formation of supercoiled loops that are cleaved at one of many potential sites. The enzyme is not released from the DNA, and ATP hydrolysis occurs presumably to provide energy for the translocation (Yuan *et al.*, 1975; Bickle *et al.*, 1978; Yuan *et al.*, 1980; Burckhardt *et al.*, 1981; Suri *et al.*, 1984; Vovis *et al.*, 1974; Yuan, 1981).

It is clear then that, in contrast to the type II and *dam* methylases, the type I methylases have a strong preference for hemimethylated DNA. Moreover, AdoMet has a dual role: it acts as a methyl donor and also as an allosteric effector for type I methylase binding to DNA. ATP is also an allosteric effector for methylase action, increasing the rate of reaction about eightfold. It can be substituted for by the nonhydrolysable $\beta\gamma$ imidoanalogue. The K_m value for AdoMet is about 0.2 μM (in the presence or absence of ATP).

By using a strain of *E. coli* K12, which has the *hsd*M and *hsd*S genes under the control of a strong phage lambda promoter, methylase can be isolated lacking the *hsd*R gene product. This methylase cannot carry out restriction (reaction 3 on previous page). With unmodified DNA, the methylation reaction is faster than with the complete restriction enzyme and is independent of ATP. Nonetheless, reactions only go to completion after incubations of about 10 hours (Vovis *et al.*, 1974; Suri *et al.*, 1984). In contrast, with hemimethylated DNA, the reaction goes to completion within 20 minutes and, once again, it is independent of ATP. It is as

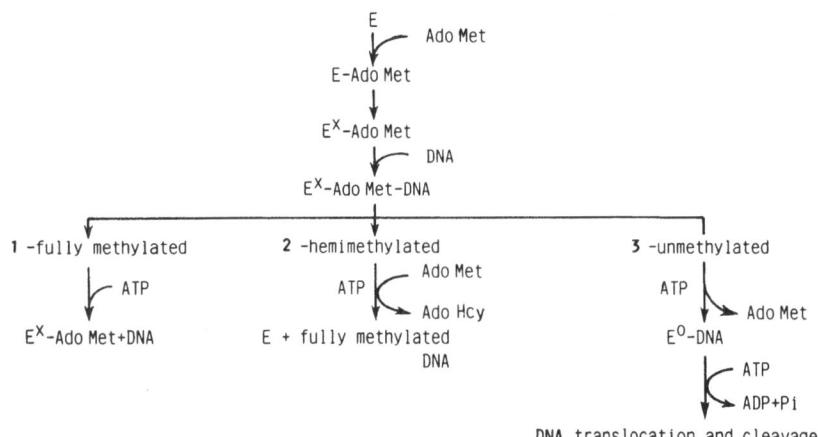

Figure 5.4. Reaction sequence for type I restriction modification enzymes. The activated enzyme-AdoMet-DNA complex follows one of three pathways depending on the state of methylation of the DNA. See text for further details. (From Burckhardt *et al.*, 1981.)

though the ATP is necessary with the complete enzyme to counter the effect of the presence of the *hsd*R coded subunit.

The *Eco*A type I restriction enzyme differs from the *Eco*B and *Eco*K enzymes in that a two-subunit methylase is all that is normally detectable in cells. However, in the presence of unmethylated DNA, a third subunit becomes associated with the complex, which now has endonuclease activity (Suri *et al.*, 1984).

5.1.4 Type III Methyltransferases (*Eco*P1, *Eco*P15, and *Hinf*III)

In some ways, the type III methyltransferases resemble the type I methyltransferases except that the *hsd*M and *hsd*S gene products are combined in a single subunit as they are in the type II methylases. ATP is not required for methylase activity (though it does stimulate it slightly). Neither is there extensive ATP hydrolysis following endonuclease action as with the type I enzymes. Both the methylase and the nuclease activity require the presence of Mg^{2+} ions (Brockes *et al.*, 1972; Reiser & Yuan, 1977; Hadi *et al.*, 1983; Iida *et al.*, 1983; Kauc & Piekarowicz, 1978).

The *Eco*P1 and *Eco*P15 restriction enzymes are encoded by almost identical DNA sequences carried on the P1 phage and the P15B plasmid respectively. Each is encoded by two genes: *res*, which is required for restriction only, and *mod*, which is required for restriction and modification. The two methylases differ only at the N-terminus. However, as with the contiguous type II genes, the type III genes appear to be transcribed from separate promoters (Iida *et al.*, 1983).

The type III restriction enzymes and the separate modification enzymes have been purified using phosphocellulose, DEAE cellulose, and DNA or heparin agarose affinity separations. The product of the *res* gene has a molecular weight of 106,000 to 110,000, and the methylase is somewhat smaller at 73,000 to 80,000. The methylase acts as a monomer, but the complete restriction enzyme has a molecular weight of 318,000 and so may contain three methylase molecules (Hadi *et al.*, 1983). The enzymes bind AdoMet very firmly. With the complete enzyme, methylation and cleavage occur together and so cleavage is never complete. All three type III enzymes studied methylate adenine residues in their nonsymmetrical recognition sequences, *i.e.*, they methylate one strand only of the DNA, which must lead to production of unmethylated duplex DNA following replication. The endonuclease makes staggered cuts 20 to 30 bases from the 3' end of the recognition site. The major factor affecting whether a type III enzyme will modify or cleave DNA appears to be the concentration of AdoMet. The K_m for the endonuclease reactions is 10^{-8} M compared with 4×10^{-7} M for methylation (Reiser & Yuan, 1977).

5.2 Mammalian DNA Methyltransferases[3]

5.2.1 The Reaction Product

5-Methylcytosine is the only minor methylated base present in the DNA of higher eukaryotes, and DNA methylases present in eukaryotic cells transfer methyl groups only to the 5 position of cytosines in DNA. Such analyses involve purifying the DNA labeled *in vitro* when the tritiated methyl group of [^{3}H-methyl]AdoMet is transferred to DNA. The DNA is then hydrolyzed to the bases (usually with formic acid, see Appendix), and these are fractionated by chromatography, electrophoresis, or HPLC.

In mammals, methylcytosine is found predominantly, if not exclusively, in the dinucleotide CpG. Hence, it was expected that the enzymes would methylate only such cytosines *in vitro*. However, Simon *et al.* (1980) concluded that the rat liver methylase would also methylate cytosines in the dinucleotides CpA and CpT–indeed methyl groups in CpA and CpT represented a major fraction of the methyl groups incorporated. A similar result has been reported for the bovine thymus enzyme (Sano *et al.*, 1983). On the other hand, there is no evidence for such methylations with the mouse enzyme. For instance when SV40, pBR322, or ϕX174 RF DNA are methylated with the mouse enzyme and then cleaved into fragments by restriction nucleases, the incorporation is proportional to the number of CpG dinucleotides per fragment (Adams *et al.*, 1984; Bestor & Ingram, 1983). Similar, more general, conclusions have been obtained by comparing the efficiency of different DNA's as substrate (Section 5.2.5).

Nearest neighbor analyses also indicate mCpG as the only product of mouse methylase action. DNA is nick translated using *E. coli* DNA polymerase I and four deoxyribonucleoside triphosphates. In four separate reactions one of the triphosphates is replaced in turn with an [α^{32}P]dNTP. When these four ^{32}P-labeled substrates are methylated *in vitro* using [^{3}H-methyl]AdoMet and the mouse DNA methylase, and then cleaved to 3' monophosphates, the methylCMP contains ^{32}P originating predominantly from dGTP, as shown in Figure 5.5.

5.2.2 Enzyme Requirements and Assay Conditions

Eukaryotic DNA methylases have no cofactor requirements and are not stimulated by ATP or Mg^{2+}. In fact, EDTA is slightly stimulatory, proba-

---

[3] Adams & Burdon, 1983.

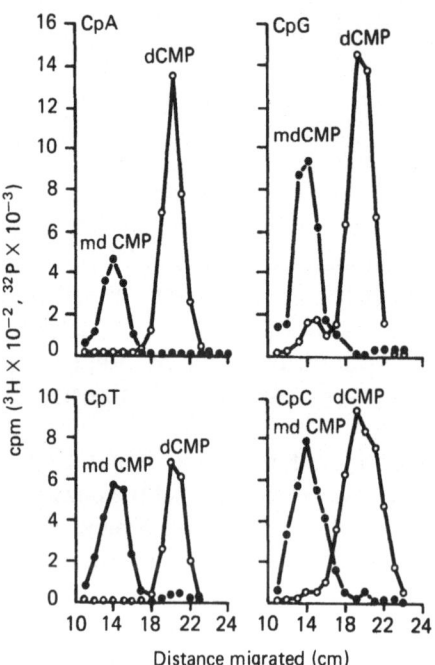

Figure 5.5. DNA from *M. luteus* was nick translated in four separate reactions, each containing a different α^{32}P-dNTP. The DNA was methylated *in vitro* using mouse DNA methylase and then hydrolyzed to the 3'dNMPs, which were separated by high voltage electrophoresis. The charts show the methyldCMP (mdCMP) and dCMP spots from the four reactions and indicate that most of the methylation is occurring in the CpG dinucleotide with a smaller amount at CpC. Open circles: ^{32}P. Closed circles: ^{3}H.

bly as a result of its ability to counter the inhibitory action of certain metal ions (Adams & Burdon, 1983). Dithiothreitol is similarly required in the reaction, which is strongly inhibited by N-ethylmaleimide (see Section 5.4). The pH optimum is between 7.5 and 8 and the K_m for AdoMet is about 2 μM (Simon *et al.*, 1978), as shown in Figure 5.6. The K_m for DNA substrate differs depending on the nature of the DNA. The range is 0.5 μg per ml to 70 μg per ml and probably reflects the number of unmethylated CpG dinucleotides present in the DNA (see Section 5.2.5). When native DNA is used as substrate, the reaction is strongly inhibited by salt—50 mM NaCl brings about a 70% inhibition. Whereas, with denatured DNA as substrate, salt is stimulatory up to 100 mM NaCl.

Particularly in crude cell extracts, the estimation of DNA methylase activity is complicated by the presence of other enzymes that transfer methyl groups to RNA and protein. It is, therefore, common practice to include in the assay procedure steps to remove protein (protease treatment and phenolization) and RNA (RNase or alkali treatment) prior to precipitation of the DNA with ethanol or acid (Adams & Burdon, 1983). The DNA may be finally immobilized on nitrocellulose or DE81 filters when the excess tritiated AdoMet is removed with saline citrate or sodium phosphate, but more usually it is simply acid precipitated onto paper filters. Measurements of methylase activity that involve monitoring the resistance of the DNA substrate to particular restriction endonucleases, (*e.g.*, HpaII) are not usual, as these only detect a small proportion of the

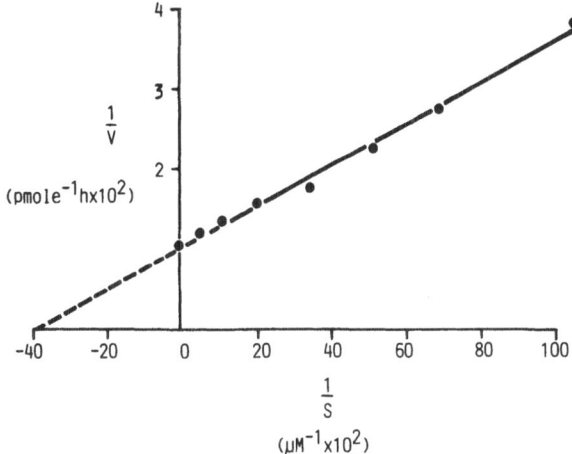

Figure 5.6. Double reciprocal plot of [AdoMet] versus methylase activity indicates a K_m value of 2.5 μM for AdoMet for the mouse ascites enzyme.

added methyl groups and in any event require the enzyme to be purified free of contaminating endonucleases (see Figure 5.1).

AdoMet at 12 μM can also act as a methyl donor in nonenzymic reactions, but this leads to synthesis of 2-methylguanine, O^6-methylguanine, and 3-methyladenine, which are not found as products of the methylase reaction. Although this is not a major problem it seems advisable to keep the concentration of AdoMet in the assays as low as practiceable (Barrows & Magee, 1982; Rydberg & Lindahl, 1982).

5.2.3   Purification and Structure

Mammalian DNA methylase has been purified from a number of tissues, as shown in Table 5.3. Except where thymus is the starting material

Table 5.3.   Vertebrate DNA Methylases Purified

Rat spleen	Kalousek & Morris, 1969b
Rat liver (regenerating)	Morris & Pih, 1971; Simon *et al.*, 1978; Ruchirawat *et al.*, 1984
Rat hepatoma	Sneider *et al.*, 1975
HeLa cells	Roy & Weissbach, 1975
Mouse ascites cells	Turnbull & Adams, 1976; Adams *et al.*, 1979a
Bovine thymus	Sano *et al.*, 1983
Mouse Friend cells	Bestor & Ingram, 1983
Human placenta	Pfeifer *et al.*, 1983; Wang *et al.*, 1984

(when the enzyme is found in the cytoplasmic fraction), activity is extracted from isolated nuclei with 0.2–0.4 M NaCl and purified by gel filtration, chromatography on DEAE, phosphocellulose, or hydroxyapatite, and affinity chromatography on heparin agarose, DNA agarose, or on a dye ligand column. A typical purification of the mouse ascites DNA methylase is given in Table 5.4.

It is very difficult to compare the specific activities of the final fractions from the various purifications, largely because of the different DNA substrates employed. This is exemplified by the human placental preparation, which shows a specific activity of over 51,000 units per mg protein when a hemimethylated DNA substrate is used but less than 5000 units using native *M. luteus* DNA (Wang *et al.*, 1984).

The molecular weight of the native enzyme has been variously estimated as between 120,000 and 280,000 and may well differ between different vertebrate tissues. SDS gel electrophoresis shows bands of 125,000 and 140,000 associated with the most pure fractions of the human placental methylase, whereas the mouse Friend cell enzyme shows bands at 150,000 and 175,000 molecular weight. The mouse ascites enzyme undergoes proteolysis in the absence of protease inhibitors, and this is associated with a decrease in the amount of the high molecular weight band on SDS gel electrophoresis and a corresponding increase in lower molecular weight bands, as shown in Figure 5.7 (Adams *et al.*, 1983b). These two forms of the native enzyme have been separated on Cibacron blue agarose (Bestor & Ingram, 1983).

In low salt, the enzyme aggregates, but it is only under these conditions that it interacts with native DNA. The reason for this is probably that the enzyme requires regions of single-stranded DNA with which to interact, and these are available when duplex DNA "breathes"—a situation that is

Table 5.4. Purification of Mouse DNA Methylase

Fraction	Total protein (mg)	Total activity (units)	Specific activity (units/mg)
Nuclear, low salt extract	5,850	374,000	64
Phosphocellulose fraction	405	216,800	535
Hydroxyapatite fraction	85.5	139,080	1,627
Gel filtration (S400)	17.8	70,440	3,959
tRNA Sepharose fraction	4.8	27,360	5,678
DNA Sepharose	0.4	4,560	11,460

Figure 5.7. SDS gel of purified mouse DNA methylase. The marker protein is α_2 macro-globulin.

—170K

encouraged at low salt and high temperature (Adams *et al.*, 1979a). It is therefore not clear whether the active form of the enzyme *in vivo* is the monomer or some multimer.

5.2.4   Cellular Location and Cell Cycle Variation

Apart from the bovine thymus (Sano *et al.*, 1983), the mammalian DNA methylase occurs associated with cell nuclei (the bovine enzyme also differs in a number of other properties). The bulk of the enzyme is, however, only loosely associated with the nuclei and can be removed by extraction with low salt treatments. Brief treatment of nuclei with micrococcal nuclease releases both oligonucleosomes and DNA methylase into the soluble fraction. On further digestion or on addition of 8 mM Mg^{2+}, most of the remaining DNA is reprecipitated, but the enzyme remains in the supernatant fraction (Adams *et al.*, 1977; Creusot & Christman, 1981), implying that the DNA methylase is associated with linker regions of chromatin.

Little DNA methylase activity is associated with transcribing regions of chromatin when these are isolated by treatment of nuclei with DNase II and Mg^{2+} precipitation of inactive regions. The enzyme is, however, preferentially solubilized along with the transcriptionally active chromatin when the Sanders (1978) procedure is used, although perhaps this is not surprising in light of the salt concentrations employed (Davis et al., in preparation).

In addition to the loosely associated enzyme, a small proportion of DNA methylase activity remains associated with nuclei even after extraction with 2 M NaCl (Burdon et al., 1985). This matrixbound enzyme is responsible for the incorporation of methyl groups into DNA when isolated nuclei are incubated with AdoMet (Davis et al., 1985; see Section 2.5). It may also be the actual form of the enzyme responsible in vivo for the maintenance methylation in the cell. The readily solubilized enzyme, on the other hand, may represent a supply of enzyme waiting for matrix attachment sites to become available (Burdon et al., 1985). When solubilized, the loosely and tightly bound forms of DNA methylase are identical in properties. Noguchi et al. (1983) have found DNA methylase activity associated with a high molecular weight complex released from nuclei by sonication. This complex (the replitase) is also associated with DNA polymerase, thymidine kinase, topoisomerase, and ribonucleotide reductase and is thought to be involved at the replication fork in organizing DNA replication.

A survey of the amount of DNA methylase extractable from various tissues indicates that the enzyme is present at only low levels in nondividing cells (Adams et al., 1974; Lutz et al., 1972). It is, therefore, not surprising that the various purifications involve the use of dividing tissues. An analysis of the specific activity of enzyme extracted from nuclei of synchronized cells reveals a sharp increase as the cells enter S phase as shown in Figure 5.8. This response is typical of enzymes involved in the synthesis of DNA and its precursors.

5.2.5   Mechanism of Action

The higher the $G+C$ content of the DNA substrate, the higher will be the number of CpG dinucleotides, and in general the better the accepting qualities of the DNA in a methylase assay. Thus M. luteus DNA (72 mole percent $G+C$) is a better accepter than is E. coli DNA (50 mole percent $G+C$). These DNA's are unmethylated at CpGs, but even mammalian DNAs will accept methyl groups in vitro, indicating that in vivo the enzymes do not seem to methylate all available sites (Sneider et al., 1975; Roy & Weissbach, 1975; Turnbull & Adams, 1976; Gruenbaum et al., 1981b). The number of CpGs that remain unmethylated may affect the K_m values for DNA, which are variously reported as between 0.5 and 2.5 μg per ml for native unmethylated DNA's but up to 30 to 70 μg per ml for native DNA from higher eukaryotes.

Figure 5.8. DNA methylase activity during the cell cycle. Confluent mouse L929 cells (S) were subcultured into medium containing aminopterin. Sixteen (16) hours later (zero time) thymidine was added to allow a synchronous wave of DNA synthesis lasting for 6 to 8 hours. DNA methylase activity was assayed in nuclear extracts from cells harvested at the indicated times.

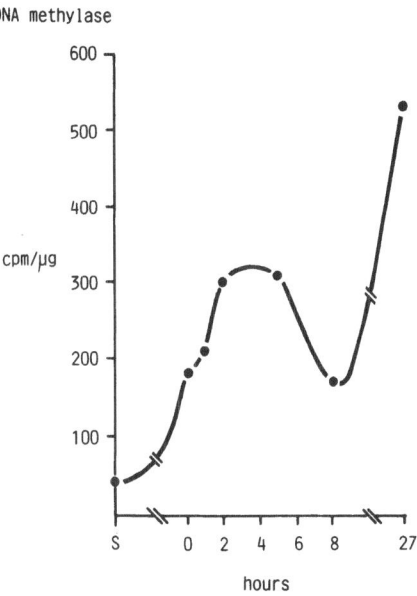

The enzyme is usually assayed using denatured DNA in a reaction that is stimulated by 100 mM NaCl. Under these conditions, greater activity is seen than when native DNA is used. In the latter case, salt strongly inhibits the binding of the enzyme, which apparently occurs at single-stranded regions in duplex or supercoiled DNA (Drahovsky & Morris, 1971; Adams *et al.*, 1979a; Adams *et al.*, 1984; Simon *et al.*, 1978).

Once bound to native, unmethylated DNA, the vertebrate DNA methylase is less sensitive to salt, and it has been proposed that the enzyme processes along the DNA in a random walk manner, adding methyl groups as it comes to unmethylated CpGs (Drahovsky & Morris, 1971). This one-dimensional diffusion, or sliding, would greatly increase the chances of the methylase finding its recognition sequence (Berg *et al.*, 1982). Inability to detect procession using methylated vertebrate DNA's as substrate may mean that the enzyme leaves the DNA on encountering a methylated region. Using an excess of cyclic ϕX174RF molecules as substrate, Adams *et al.* (1984) could find only partial support for the processive mechanism, implying that even on unmethylated duplex DNA, the methylase will only process for a short distance.

There is not much evidence for *de novo* methylation occurring in somatic cells, and hence the reaction catalyzed by DNA methylase using unmethylated DNA as a substrate may not reflect the *in vivo* situation. Indeed, the inability of the enzyme, *in vitro*, to bind to duplex DNA except at single-stranded regions may explain the low frequency of *de novo* methylations found *in vivo*. Alternatively, the DNA methylase may have a domain that will recognize a methylcytosine on one strand of hemimethylated DNA, and only then will the active site catalyze the transfer of a methyl group to the unmethylated strand. Such a mechanism could explain the increase in activity on unmethylated DNA following limited proteolysis of the enzyme (Adams *et al.*, 1983b). Removal of the domain that recognizes the hemimethylated site may relieve a constraint, thereby allowing the enzyme to work on totally unmethylated duplex DNA, as shown in Figure 5.9. It is surprising that the vertebrate DNA methylases are as large as 150,000 to 180,000 molecular weight, but this may be partially explained if they have such domains important in control.

Nonetheless, the major function *in vivo* of DNA methylases is the maintenance of the pattern of methylation (see Section 2.3). The natural substrate of DNA methylases is the hemimethylated DNA that results following DNA replication, and, *in vitro*, the enzyme shows much greater

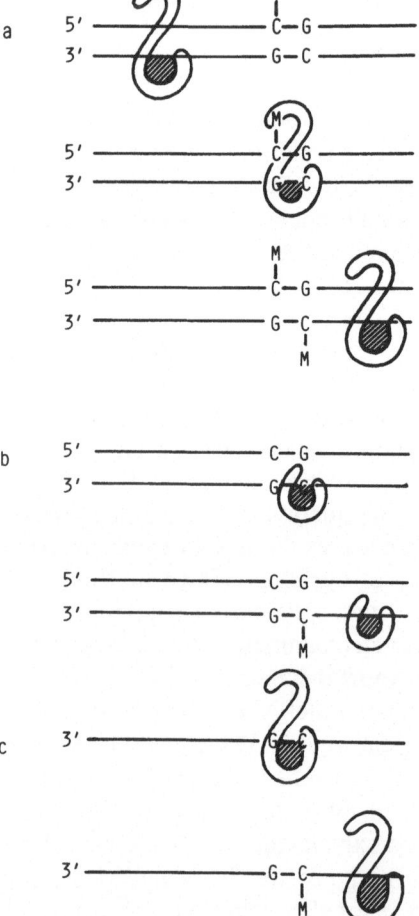

Figure 5.9. A model of the postulated domain structure of mouse DNA methylase. (a) on hemimethylated DNA a region of the enzyme recognizes the methyl group on one strand, and a catalytic region adds a methyl group to the cytosine on the other strand. (b) the proteolized enzyme's catalytic unit will add methyl groups *de novo*. (c) with single-stranded DNA the recognition region is not involved.

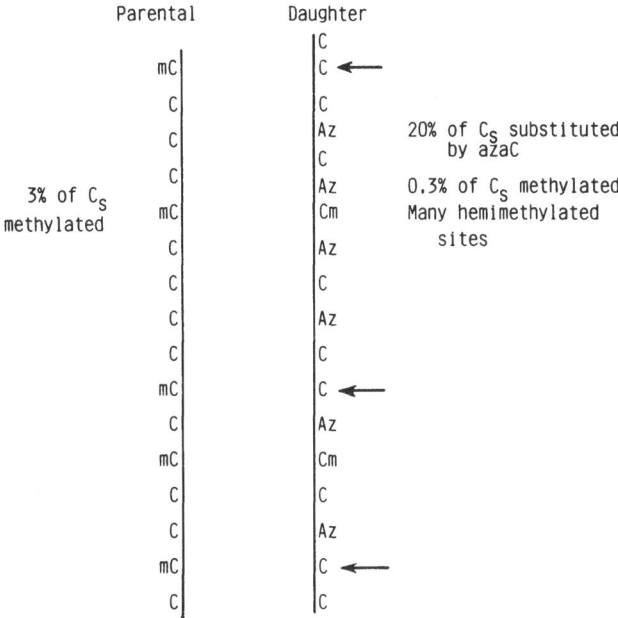

Figure 5.10. DNA from cells treated with 5-azacytidine. The parental strand has about 3% of the cytosines methylated. The daughter strand, which can have up to 20% of the cytosines replaced by azacytosine, has only 0.3% of the *remaining* cytosines methylated, thus leaving many hemimethylated sites.

activity with hemimethylated DNA than with native or denatured DNA (Adams *et al.,* 1979a; Jones & Taylor, 1981; Taylor & Jones, 1982; Adams *et al.,* 1982; Gruenbaum *et al.,* 1982; Pfeifer *et al.,* 1983; Bestor & Ingram, 1983; Wang *et al.,* 1984).

There are several different ways to prepare hemimethylated DNA. One of the simplest is to prepare DNA from cells that have been treated with azacytidine or azadeoxycytidine for the duration of one S phase. As described in Section 5.4, such treatment leads to the inactivation of the DNA methylase, and the newly synthesized strand of DNA is deficient in methyl groups, as shown in Figure 5.10. One disadvantage of this substrate is that, in addition to having hemimethylated CpG dinucleotides, it also contains a number of azacytosine residues that may or may not have undergone ring cleavage. Intact azacytosine bases on the 5′ side of guanines will interact and inactivate DNA methylase *in vitro* (see Section 5.4.3). Hence, it is very important to have the extent of substitution at the optimal level to give the maximum number of hemimethylated sites with the minimum number of azacytosine substitutions.

Nick translation reactions with DNA polymerase I of *E. coli* have also been used to prepare hemimethylated DNA. Starting with either calf thymus DNA or DNA from phage XP12 (which has all of its cytosines replaced with methylcytosine), sections of the DNA can be replaced with unmethylated DNA, as shown in Figure 5.11. Alternatively, unmethylated duplex DNA can be nick translated in the presence of methyl dCTP, or single-stranded phage DNA can be converted to the duplex form in the presence of methyl dCTP. Double-stranded poly (dG, dC, dm⁵C) has been prepared in which 32% of the CpG dinucleotide pairs are hemimethylated (Bestor & Ingram, 1983).

Using the mouse DNA methylase, Gruenbaum *et al.* (1982) obtained virtually no methylation with single-stranded or duplex φX174 DNA. This observation is difficult to understand because other groups have shown that denatured DNA is a good substrate for this enzyme. However, they obtained high levels of incorporation into hemimethylated DNA. Wang *et al.* (1984) have shown that the relative incorporation into duplex and hemimethylated DNA depends critically on the conditions and the length of the incubation, as shown in Figure 5.12. In particular, the *de novo* reaction (*i.e.*, methylation of unmethylated DNA) appears to show a short

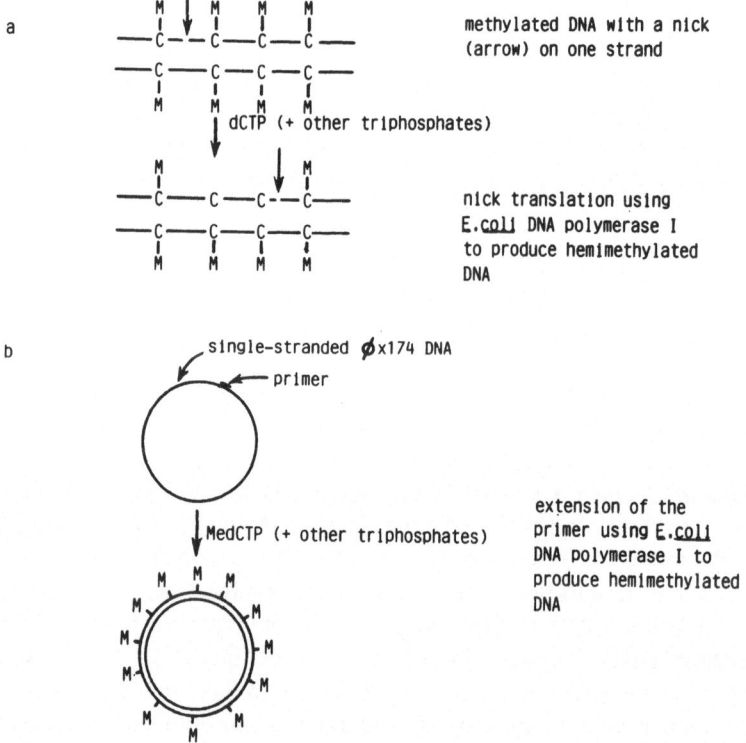

Figure 5.11. Two other ways of preparing hemimethylated DNA. (a) vertebrate DNA, which is highly methylated, can be nick translated using dCTP and the other deoxyribonucleoside triphosphates. (b) single-stranded unmethylated phage DNA can be converted to the duplex form using methyl dCTP in place of dCTP.

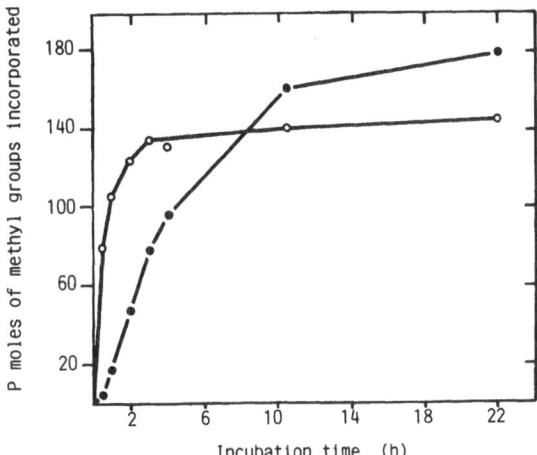

Figure 5.12. Kinetics of *de novo* and maintenance methylation. The time course is shown of methylation of hemimethylated XP12 DNA (○) or of unmethylated *M. luteus* DNA (●) using the placental methylase. (From Wang *et al.*, 1984.)

lag, which may be related to the proteolytic activation of the enzyme or the slow binding of the enzyme to "breathing" sites.

In general, as with prokaryotic type I methylase, the *de novo* methylation reactions only go to completion after very long incubations with high ratios of enzyme to DNA (up to 10,000 units per μg). The vertebrate methylases also resemble the type I methylases in their preference for hemimethylated substrates, and in this case reactions proceed far more rapidly.

5.3 DNA Methyltransferases in Other Eukaryotes

Although DNA methylase has been measured in *Xenopus laevis* (Adams *et al.*, 1981b), fat head minnow cells (Willis *et al.*, 1984), mosquito cells (Adams *et al.*, 1979b), and pea seedlings (Kalousek & Morris, 1969a), it has only been purified and studied in detail in *Chlamydomonas reinhardi* (Sano & Sager, 1980).

The frog and fish DNA methylases seem to resemble the mammalian enzymes in sizes, cellular location, and substrate preferences. Unusually, frog virus 3 encodes a new DNA methylase found in the cytoplasm of infected cells (Willis *et al.*, 1984). This enzyme is unusual in that it prefers native unmethylated DNA to hemimethylated DNA.

In mosquito cells a small amount of DNA methylase is detected in the cytosol, and this transfers methyl groups to DNA adenine as well as to DNA cytosine.

The enzyme from pea seedlings shows maximal activity at 30°, and there is no preference for a single-stranded DNA substrate (Dolan & Adams, unpublished data.)

The *Chlamydomonas* enzyme has been purified 900-fold from vegetative cells and is the smallest eukaryotic methylase so far investigated, having a native molecular weight of only 55,000. It requires no cofactors, but is stabilized by dithiothreitol. It is remarkable in that a native DNA substrate accepts ten times more methyl groups than the corresponding denatured DNA substrate; *i.e., de novo* activity seems to predominate as with the frog virus 3 enzyme. Homologous nuclear DNA is a good substrate for the enzyme, as is the heavily methylated calf thymus DNA. Only about half the methyl groups introduced are in the dinucleotide mCpG. As DNA of vegetative *Chlamydomonas* cells lacks methylcytosine (Chapter 11), the finding of this enzyme was a surprise. However, female gametes and zygotes have a 200,000 molecular weight DNA methylase that is postulated to be a multimer or a high molecular weight active form of methylase (Sager *et al.,* 1984). The active enzyme introduces methyl groups into CpA, CpG, CpT, and CpC dinucleotides, but does not methylate poly(dC)·poly(dG). It is possible that the enzyme methylates the sequence CNG (where N is any nucleotide), as this is the sequence methylated in higher plant DNA. This form of the *Chlamydomonas* enzyme shows maximum velocity with a hemimethylated substrate and so resembles the mammalian enzyme. It is presumably responsible for the maintenance of methylation patterns in female gametes and zygotes. A second enzyme may exist with a specificity similar to the mammalian enzymes, as chloroplast DNA is partly resistant to HpaII restriction, though this may simply be a consequence of the fact that the two cytosines in the recognition site (CCGG) both occur in the sequence CNG.

5.4 Inhibitors of DNA Methylation

No agent has been described that uniquely inhibits the action of DNA methyltransferase. Mouse DNA methylase activity *in vitro* is blocked by reagents that react with -SH groups, *e.g.,* N-ethylmaleimide or p-chloromercuribenzoate; by substrate analogues; and by reagents that interfere with the interaction of the enzyme with its substrates, *e.g.,* detergents, salt, and heparin, all of which reagents will interfere with many other reactions. Cordycepin (3′ deoxyadenosine) following conversion to 3′-d AdoMet inhibits methylation of *both* RNA and DNA in human lymphoblasts (Kredich, 1980), and this is typical of analogues of AdoMet. As the enzyme binds to but does not methylate phage the DNA or RNA these also inhibit its action (Bolden *et al.,* 1984).

5.4.1 S-Adenosylhomocysteine and Ethionine

As described in Chapter 4, S-adenosylmethionine (AdoMet) serves as a methyl donor in the transfer of methyl groups not only to DNA but also to RNA (Perry & Kelley, 1974; Maden & Salim, 1974; Kerr & Borek, 1972), proteins (Paik & Kim, 1975), and many low molecular weight compounds (Greenberg, 1963). Besides the methylated macromolecule, the other major product of the reaction is S-adenosylhomocysteine (see Figure 2.1), which is a potent inhibitor of all these transmethylation reactions with K_i values ranging from 0.3 to 20 μM. The inhibition of mouse DNA methylase *in vitro* by AdoHcy is shown in Figure 5.13.

In vivo the AdoHcy produced is degraded by a specific hydrolase to homocysteine and adenosine, and this is deaminated to inosine and then to hypoxanthine, which is either reincorporated or excreted as uric acid (see Figure 4.5). Growth of cells is inhibited in the presence of adenosine (30–50 μM), and this effect is accentuated by addition of homocysteine thiolactone, which leads to elevated intracellular pools of S-adenosylhomocysteine (Kredich & Hershfeld, 1979; Kredich & Martin, 1977). Similarly, the addition of an inhibitor of adenosine deaminase (erythro-9-(2-hydroxy-3-nonyl)adenine, or EHNA) produces increased pool sizes of AdoHcy. Consequently, intracellular methylation reactions are blocked, though which particular one is essential for cellular survival is not clear. Figure 5.14 shows the effect of the increased intracellular AdoHcy concentrations on DNA methylation.

Injections of ethionine into rats or incubation of sea urchin embryos in medium containing ethionine leads to inhibition of DNA synthesis, but there is some controversy over whether or not the DNA that is made is

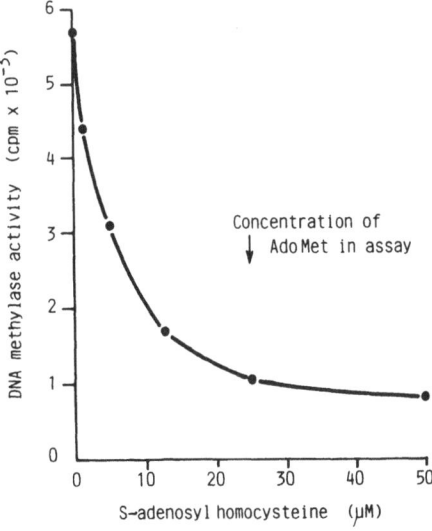

Figure 5.13. Inhibition of DNA methylation *in vitro* by AdoHcy. The inhibitor was added to the assay for mouse DNA methylase that contained the substrate (AdoMet) at 24 μM. Reprinted with permission from Adams and Burdon, *Enzymes of Nucleic Acid Synthesis and Modification*, p. 135. Copyright 1983, CRC Press, Inc., Boca Raton, Florida.

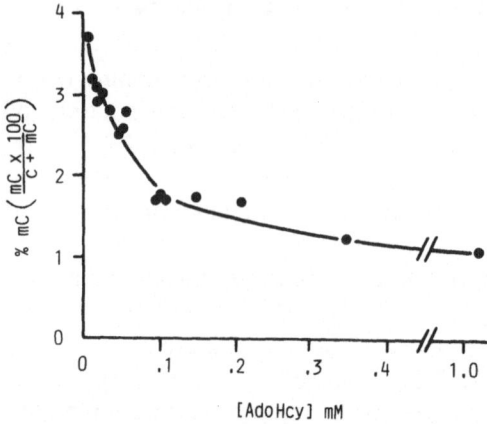

Figure 5.14. Inhibition of DNA methylation *in vivo* by AdoHcy. The intracellular concentration of AdoHcy was manipulated in mouse lymphoma cells by treatment with adenosine and L-homocysteine thiolactone. DNA was measured following a 2-hour incubation with 6-^{3}H-uridine as described in Section A.3.1. The figure is based on data in the paper of Kredich and Martin, 1977.

undermethylated (Cox & Irving, 1977; Graziani *et al.*, (1970). Incubations of cells in culture with ethionine or cycloleucine at low concentrations has little inhibitory effect on DNA synthesis, but may cause a slight reduction in levels of DNA methylation (Jones & Taylor, 1981; Woodcock *et al.*, 1983). Any effects of these reagents are, however, unlikely to be specific to methylation of DNA, as they will presumably affect most methyltransferases in a similar manner. Sneider *et al.* (1975) have shown that S-adenosyl ethionine inhibits DNA methylase activity.

5.4.2 Alkylating Agents

Alkylating (and arylating) agents interfere with methylation *in vitro*, acting both on the DNA substrate and on the enzyme (Salas *et al.*, 1979; Cox, 1980; Wilson & Jones, 1983). Nitrogen mustard is one of the most potent inhibitors, and this can also act as a cross-linking reagent, although the bulky nature of the N-acetoxy-N-acetylaminofluorene and similar adducts may account for their potent inhibitory action.

N-2-Acetylaminofluorene and its metabolite N-acetoxy-N-acetylaminofluorene bind irreversibly to the C8 of guanine residues in DNA and induce conformational changes nearby because the fluorene ring intercalates into the DNA leading to some local denaturation (Fuchs & Danne, 1972). Such DNA is a poor substrate for DNA methylase, but is bound to the enzyme more tightly than nonsubstituted DNA, indicating that its action may be to inhibit movement of the enzyme along the DNA

(Pfohl-Leszkowicz *et al.*, 1981; Ruchirawat *et al.*, 1984). In contrast, DNA modified with an aminoquinoline N-oxide group on C8 of guanine residues accepts more methyl groups than control DNA and shows a reduced affinity for the enzyme (Pfohl-Leszkowicz *et al.*, 1983).

5.4.3 5-Azacytidine and 5-Azadeoxycytidine (Figure 5.15)

Interest in these drug centers on their potential as inhibitors of *in vivo* methylation. It is perhaps wise, however, to look first at the basic metabolism of the drugs and some of the more drastic effects they may produce on other cellular processes.

On entering the cell these drugs are rapidly phosphorylated (an exception is *B. subtilis*—Guha, 1984). Azacytidine is phosphorylated by uridine/cytidine kinase, for which it competes with uridine and cytidine (Lee *et al.*, 1974). Azadeoxycytidine and deoxycytidine are phosphorylated by deoxycytidine kinase (Momparler & Derse, 1979).

5-Azacytidine is subject to deamination to 5-azauridine, and either of these can be phosphorylated and incorporated into RNA. The former may also be incorporated into DNA to a lesser extent. The deoxyanalogue is incorporated largely into DNA in eukaryotes, though in bacteria, deamination is followed by conversion to 5-azauracil and incorporation into RNA (Cihak & Vesely, 1977; Vesely & Cihak, 1977, 1978; Adams *et al.*, 1982; Taylor *et al.*, 1984).

The azanucleosides are unstable in aqueous solutions and rapidly break down by ring-cleavage reactions. Perhaps because of this, most of the *in vivo* action of the drugs depends on their incorporation into nucleic acids (where they appear to be more stable), rather than on the effects of the nucleosides and nucleotides on other metabolic conversions. Among the latter are the inhibition of orotidine 5'-phosphate decarboxylase by azacytidine phosphate (Cihak & Broucek, 1972), and of thymidylate synthetase by azadeoxyuridine (Vesely *et al.*, 1969).

Azacytidine is incorporated into all species of RNA and causes a rapid inhibition of protein synthesis and breakdown of polysomes (Cihak *et al.*,

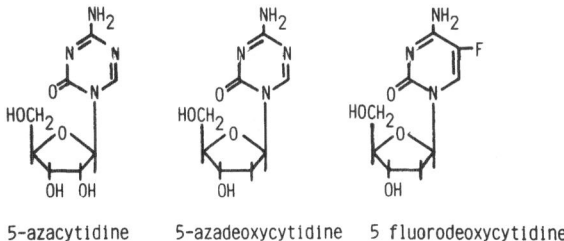

5-azacytidine   5-azadeoxycytidine   5 fluorodeoxycytidine

Figure 5.15. Structure of three 5-substituted analogues of deoxycytidine.

1973; Zain et al., 1973; Reichman & Penman, 1973). This leads to a reduced rate of breakdown of some enzymes (Cihak & Vesely, 1973; Cihak & Kroger, 1981; Cihak et al., 1972a; Levitan & Webb, 1969) and enhanced hormonal induction of tryptophan pyrrolase in rat liver (Levitan & Webb, 1969). Injections of azacytidine into rats 30 minutes after partial hepatectomy abolish DNA synthesis 24 hours later, and similar inhibitory effects are noted with phytohemagglutin stimulated lymphocytes (Cihak et al., 1972a; Cihak, Seifertova, Vesely & Storm, 1972; Zain et al., 1973). To obtain these effects, only short exposure to the drug is required at a time long before the time of initiation of DNA synthesis, implying that the effect is mediated via incorporation into RNA. The deoxyanalogue does not produce these effects (Vesely & Cihak, 1978).

These inhibitory effects may be the result of blocking methylation of the precursors to ribosomal RNA (Weiss & Pitot, 1974) or transfer RNA (Lu et al., 1976; Lee & Karon, 1976). It is probable that the presence of a small amount of RNA containing azacytosine is sufficient to inactivate the RNA cytosine methyltransferases whose action is essential for correct maturation of the RNA. A similar mechanism accounts for the inhibition of DNA cytosine methyltransferases (see below).

Azacytidine and azadeoxycytidine kill cells after a lag of several hours as the result of their incorporation into DNA (Li et al., 1970; Zain et al., 1973; Adams et al., 1982).

Although azacytosine can be recovered intact from the DNA of cells treated with azacytidine (Jones & Taylor, 1981), it is likely that some of the deleterious effects of the analogues are caused by breakdown of the triazine ring. Chromosome damage can be seen (Karon & Benedict, 1972; Li et al., 1970; Harrison et al., 1983), and DNA isolated from treated cells has a much lower molecular weight than controls (Zadrazil et al., 1965) as shown in Figure 5.16.

On incorporation into DNA, azacytosine brings about an inhibition of DNA methylation. Although the triazine ring itself cannot accept methyl groups on the 5 position, this is not the only reason for the inhibition. The extent of inhibition is much greater than the level of incorporation of azacytosine would lead one to expect. Furthermore, gross inhibition of methylation of cytosine residues occurs in DNA containing only a few azacytosines (Jones & Taylor, 1980; Adams et al., 1982; Creusot et al., 1982; Wilson et al., 1983).

The result of this inhibition is to produce hemimethylated DNA in which the daughter strand of DNA synthesized in the presence of the drug has up to 20% of the cytosines replaced by azacytosine (presumably at random), but fewer than 10% as many methylcytosines as the parental strand (see Figure 5.10). Because of the presence of hemimethylated sites, such DNA might be expected to be an excellent substrate for DNA methyltransferase. This is found to be true but only when cells are treated with low doses of azacytidine or azadeoxycytidine ($<10^{-6}$ M) prior to

Figure 5.16. Size of DNA from control (o) and azadeoxycytidine-treated (•) cells labeled with ^{14}C deoxycytidine. DNA was isolated in parallel from control mouse L929 cells and cells treated with azadeoxycytidine (1 μM) for 10 hours. The DNA was sedimented on a 10 to 25% glycerol gradient.

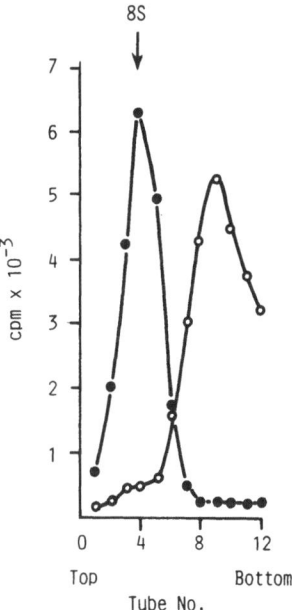

isolation of DNA. Higher doses ($10^{-5} M$) leading to higher levels of incorporation of the drug produce hemimethylated DNA, which is a poorer substrate than control DNA, as shown in Figure 5.17 (Jones & Taylor, 1981; Taylor & Jones, 1982; Adams *et al.*, 1982). Moreover, heavily substituted DNA actually inhibits methylation of other DNA's present in the enzyme incubation (Taylor & Jones, 1982; Adams *et al.*, 1984; Christman, 1984). It is as though the presence of azacytosine in DNA blocks its methylation both *in vivo* and *in vitro*.

Associated with this inhibition of DNA methylation is a disappearance of DNA methyltransferase activity from treated cells (Creusot *et al.*, 1982; Tanaka *et al.*, 1980; Taylor & Jones, 1982; Burdon *et al.*, 1985). Methyltransferase activity disappears quickly from low salt nuclear extracts of treated cells and cannot be recovered from any other fraction, though the amount of matrixbound enzyme does increase slightly (Burdon *et al.*, 1985).

These findings can be explained if DNA methyltransferase binds irreversibly to DNA containing azacytosine, and Santi *et al.* (1983, 1984) have proposed that this is highly probable in light of the mechanism of action of enzymes catalyzing electrophilic substitution at the 5 position of pyrimidines. They propose, as Figure 5.18 illustrates, that the enzyme, via an SH group, acts at the 5:6 double bond and becomes bound to the 6 position, thereby activating the otherwise inert carbon in position 5. Addition of a methyl group and β elimination release the enzyme again. Of course, the presence of a nitrogen in place of carbon in position 5 will lead

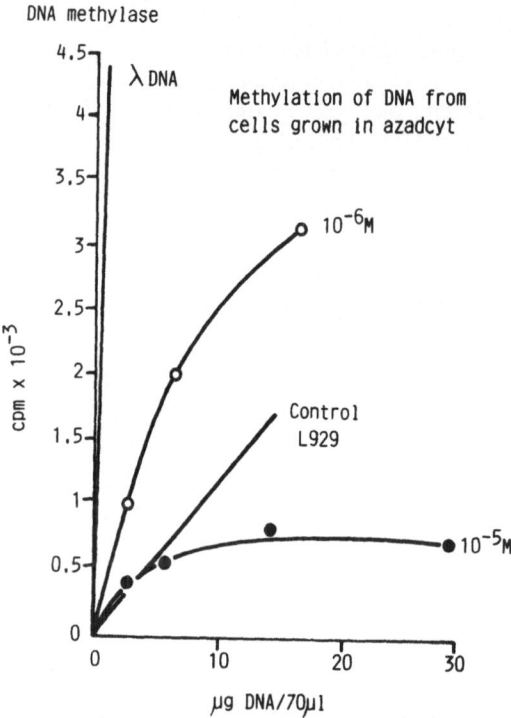

Figure 5.17. Effect of azacytidine substitution on substrate efficiency. DNA isolated from control mouse L929 cells and from cells incubated with azadeoxycytidine ($10^{-5} M$ or $10^{-6} M$) for one S phase was used as a substrate for mouse ascites DNA methylase. DNA with the lower amount of substitution is the better substrate. Incorporation using lambda DNA is shown for comparison.

to accumulation of the DNA-enzymebound intermediate, or more likely to breakdown of the triazine ring to release a formylated enzyme.

A similar mechanism can be used to explain the inhibitory effects of DNA containing 5-fluorodeoxycytidine, though the major cytotoxic action of this drug follows its deamination to 5-fluorodeoxyuridine, which inhibits thymidylate synthase (Newman & Santi, 1982). Proteins other than DNA methylase may also bind to azacytosine containing DNA (Christman et al., 1985), and formation of any stable DNA protein complex may affect gene expression.

At 10 μM azacytidine, Jones and Taylor (1981) showed the substitution of 1.3% of DNA cytosines in 20 hours. Because of the instability of the drug, the substitutions probably occurred in the first 3 to 4 hours and would amount to the incorporation of about one million molecules of azacytosine into CG dinucleotides per cell. This dose of azacytidine completely inactivates the DNA methyltransferase in about 4 hours (Creusot

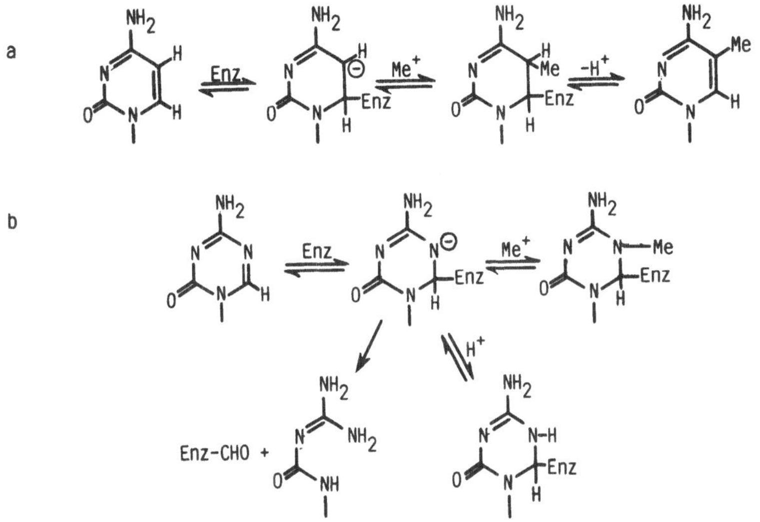

Figure 5.18. Proposed interaction of DNA methylase (Enz) with DNA. (a) shows the normal reaction where the enzyme is covalently bound to the 6 position, while the methyl group is added to the 5 position of the cytosine. In (b) the normal reaction cannot go to completion, and the enzyme substrate complex either remains intact or the formylated enzyme is released concomitantly with ring opening.

et al., 1982), but it is unlikely that there are as many as one million molecules of DNA methyltransferase per cell.

As we shall see in Chapter 8, the incorporation of 5-azacytosine into DNA leads to marked changes in metabolism and development. Whether or not these changes are a result of the inhibition of methylation or simply of the presence of an abnormal base in DNA is not yet known. It is probable that both of these actions, together with the inhibition of protein synthesis and RNA methylation play their parts in altering cellular differentiation. It is highly relevant, however, that in *Aspergillus* sp that lack methylcytosine in their DNA, azacytidine produces developmental changes as a result of changes in expression of one particular gene (Tamame *et al.*, 1983). On the other hand, we have found growth of insect cells to be unaffected by the presence of 10^{-5} *M* azacytidine, though we have no evidence in this case that the drug is incorporated into DNA (Adams R. L. P., unpublished data).

6

The Function of DNA Methylation in Bacteria and Phage

6.1 Restriction Modification

In many ways, the bacterial restriction-modification system is akin to the immune system of higher animals. It provides the means whereby an infected bacterium can repel the invading bacteriophage. When a phage attempts to grow on a new host species of bacterium, its efficiency is only a fraction of a percent of that shown on its normal host. This is a major factor in determining the host range of bacteriophage. Those few phage that do grow produce new phage that will now efficiently infect the new host species. The restriction of phage to particular host species is brought about by the presence in the host of enzymes capable of bringing about endonucleolytic degradation of foreign DNA. The DNA is recognized as being foreign because it does not contain modifications (usually methylations) on particular bases. The modifications are brought about by DNA methylases with a sequence specificity to match the endonucleases present in the same cell, shown in Figure 6.1.

In bacteria carrying a restriction-modification system, the DNA is always modified. Even immediately after replication the parental strand is modified and hence not susceptible to restriction endonucleases, which in general do not act on hemimethylated DNA. The daughter strand is methylated after only a very short time interval (Section 2.2). When an unmodified phage DNA molecule enters the bacterial cell there is a competition between the restriction endonuclease and the methylase. As methylation must be complete to protect the incoming molecule, and as one cut will probably inactivate the phage, it is clear that the system can be very efficient.

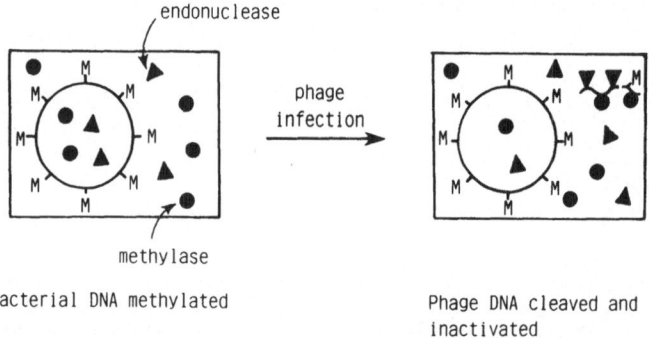

Figure 6.1. Restriction-modification. When an unmodified phage infects a bacterium carrying a restriction-modification system, the phage DNA represented as a wavy line is acted upon by the methylase (●) and endonuclease (▲). Although some modification occurs, nearly all molecules are cleaved and growth is restricted. The large circle represents the bacterial chromosome, which is methylated and so resistant to the endonuclease action.

When protection against phage-coded nucleases is afforded by *complete* replacement of a base by an unusual base (*e.g.*, hydroxymethylcytosine, Wyatt & Cohen, 1953; uracil or methylcytosine, Kuo & Tu, 1976), the modification takes place at the mononucleotide level. To this extent it resembles the modification of dUMP to dTMP by addition of a methyl group to the 5 position of the base using a folic acid coenzyme. However, in the traditional restriction-modification systems, only occasional bases are modified, and the modification occurs at the polymeric DNA level (see Section 2.4). The modifications used to protect foreign DNA against nuleolytic breakdown are varied, but we shall consider only those based on methylation—almost exclusively of the 5 position of cytosine or the N6 position of adenine (although N4 methylcytosine has also been reported by Janulaitis *et al.*, 1983). Each modification methylase recognizes a specific sequence that may consist of a set of 4 or 6 bases, and the same sequence is recognized by the corresponding endonuclease. Thus methyl groups may be present on host DNA only once every 250 to 4000 bases and restriction enzymes will cleave unmodified DNA into fragments of a similar size. As the sequences recognized by the type II enzymes are short palindromes, the methylase produces DNA with a two methyl groups per site, one on each strand:

$$e.g., \text{GmA T C}$$
$$\text{C T AmG}$$

Restriction-modification systems are of three types. Type II is the simplest and consists of separate methylases and endonucleases. The nucleases require Mg^{2+} and the methylases AdoMet for activity, but nothing further. A type III methylase can exist alone or in conjunction with a larger protein when, as well as showing methylase activity, it can act as an endonuclease. The endonuclease requires ATP as well as Mg^{2+} for its activity, which is stimulated by AdoMet. Hence, in the presence of Ado-Met cleavage and modification are competing activities. Type I systems are even more complex and consist of three nonidentical subunits coded for by three contiguous genes on the *E. coli* chromosomes. Mg^{2+}, ATP, and AdoMet are required for restriction and modification. The restriction takes place away from the recognition site and is followed by a large amount of ATP hydrolysis as the enzyme translocates around the DNA. The details of the methylase activities of bacterial restriction-modification systems are dealt with in Chapter 5.

Hundreds of type II restriction endonucleases are now known, covering a very broad range of sequence specificities. Although few of the corresponding methylases have been described, it is assumed they exist and would be revealed by a search in the appropriate bacterium. Isochizomers (nucleases that recognize the same sequence) may cleave that sequence in different places and/or be inhibited by methylation of different bases. Thus MspI does not cleave CCGG methylated on the first cytosine and HpaII action is blocked by methylation of the second cytosine. Both these enzymes cleave the sequence between the cytosines to give DNA fragments with staggered ends

$$5'\text{-C} \qquad \text{-C G G-}3'$$
$$3'\text{-G G C} \quad + \quad \text{C-}5'$$

ApyI cleaves the sequence $CC_{T}^{A}GG$ after the two cytosines, whereas *Eco*RII acts before the two cytosines, but not if the second cytosine is methylated, a modification that does not affect ApyI (McClelland, 1981; Roberts, 1980).

Some restriction enzymes are blocked by modification at sites other than those modified by their cognate methylases. Thus TaqI methylase modifies the A residue in the sequence TCGA and this interferes with restriction not only by TaqI but also SalI (GTCGAC) and XhoI (CTCGAG), whose actions are also blocked by methylation of the internal cytosine. DpnI is an unusual type II restriction enzyme that cleaves only methylated DNA, recognizing CmATC (Lacks & Greenberg, 1977). A list of restriction enzymes and their recognition sequences can be found in Roberts (1982, 1984).

In most cases it is not known whether the nuclease and methylase are encoded on the bacterial chromosome or on plasmids carried by the bacterium. However, in *E. coli* different strains carry different type I restriction-modification systems encoded by the bacterial chromosome and some have type II systems, *e.g.*, *Eco*RI, *Eco*RII, and *Eco*RV, or type III systems, *e.g.*, *Eco*P1 and *Eco*P15 carried on plasmids or phage. *M. luteus* and *B. radiodurans* appear to lack all methylation systems, indicating that these are not an essential feature for survival (Vanyushin *et al.*, 1968; Shein *et al.*, 1972). Indeed, it is interesting to speculate that restriction-modification systems may have arisen from the mechanism used by the T-even phages to ensure their own replication and the death of the host cell. Here the phage DNA is modified (by hydroxymethylcytosine glycosylated to varying extents) and carries genes for enzymes that will (a) destroy unmodified DNA, and (b) synthesize and polymerize hydroxymethyl dCTP (see Adams *et al.*, 1981a for references). It is perhaps a short step from here to the development of the simpler methylation-restriction enzyme situation and the turning of the tables that allows such a system to be used against invading phage. Of course phage can take countermeasures, and phage T3 and T7 code for a protein that inhibits the nuclease and ATPase activity of the EcoB restriction enzyme (Mark & Studier, 1981; Spoerel *et al.*, 1979).

Restrictionless mutants are readily obtainable and many of those in type I or III systems still retain DNA methylase activity (Glover, 1970; Meselson *et al.*, 1972).

This may explain the finding in some bacteria of methylases for which no corresponding endonuclease exists (Brooks & Roberts, 1982). Yet the ability of a host to resist infection by an invading phage is increased if several different restriction-modification systems are operating (Arber & Morse, 1965; Boyer, 1971). In fact *M. bovis* and *N. gonorrhaea* do seem to contain several methylases and their DNA is resistant to cleavage by many unrelated enzymes (Brooks & Roberts, 1982). That this is not more commonplace may be due to the problems posed by the presence of excessive amounts of methylated bases in DNA (see Section 7.2).

Besides interfering with the infection of bacteria with phage DNA, the restriction enzymes can prevent conjugation and exchange of plasmids between bacteria containing different restriction enzymes. However, by cleaving the incoming DNA, the efficiency of recombination may be increased (Chang & Cohen, 1977; Boyer, 1971). Nonetheless, the presence of a particular restriction-modification system may provide a strong barrier to interspecies conjugation and may be one of the major factors in speciation.

The search for restriction-modification systems in eukaryotes has been unsuccessful, although enzymes that cleave satellite DNA's into discrete size fragments have been reported (Brown *et al.*, 1978). These may reflect preferences of nucleases for cleavage sites in the repeated sequences, as has been reported for micrococcal nuclease (Keene & Elgin, 1981). No inhibition of cleavage by methylation has been reported. In contrast, methylated DNA in chloroplasts escapes the degradation meted out to unmethylated sequences, though no specific nuclease has been characterized in this case (see Section 11.7.1).

6.1.1 Methylation of φX174 DNA

Razin *et al.* (1973) first showed that during replication of φX174 in *E. coli* C the newly synthesized viral strands of DNA become methylated, presumably by the DNA methylase present in the host bacterium. When φX174 infected the methylcytosine-deficient *E. coli* B, it appeared that a new methylase was induced, which leads to the addition of a single methyl group per φX174 genome (Razin, 1973). This methyl group was thought to be involved in the recognition site for the gene A protein, which initiates RF replication and cleaves off unit-length viral DNA molecules (Friedman & Razin, 1976; Friedman *et al.*, 1977). The methylation was implied to take place at the sequence $CC^{A}_{T}GG$, which is one of the sites recognized by the dcm methylase (see below), and which becomes resistant to cleavage by EcoRII when methylated. Two such sites are found in φX174 RF DNA, but methylation is seldom complete. However Hattman *et al.* (1979) showed that φX174 growing on methylcytosine-deficient *E. coli* C *dcm⁻* lacks methyl groups, is sensitive to EcoRII, and grows normally, indicating that methylation is not essential for activity of gene A protein. φX174 growing in *dcm⁺* cells is resistant to EcoRII.

6.2 *dcm* and *dam* Methylation

The EcoRI methylase will lead to the presence of one methyladenine about every 4000 bases, *i.e.*, 0.025 mole percent, and the EcoB enzyme will add a further 0.01 mole percent methyladenine to *E. coli* B DNA. These figures are very low and do not explain the 0.5 mole percent of methyladenine found in *E. coli* B and the similar amount of methylcytosine also present in *E. coli* C. In fact, most methyl groups are added to *E. coli* DNA by one of two methyl transferases coded for by the *dam* (DNA adenine methylase) or *dcm* (DNA cytosine methylase) genes. There are no cognate endonucleases for the *dam* and *dcm* (or *mec*) methylases, which methylate *E. coli* DNA in the sequences GATC and $CC^{A}_{T}GG$, respectively. The *dcm* methylase was also thought to methylate other sequences (Razin *et al.*, 1980), but it now appears that $CC^{A}_{T}GG$ sites are present in *E. coli* DNA more often than expected, and the observed frequency of such sites is in agreement with the amount of methylcytosine present (Szyf *et al.*, 1982). It is possible that there were endonucleases corresponding to these methylases, but that they lost their activity and can no longer be detected. Restriction endonucleases do exist that recognize the unmodified sequence, and the EcoRII enzyme, which cleaves CC^A_TGG, can be present in the same cells as the *dcm* methylase, but is carried on a plasmid, and the two are inherited independently. Alternatively, the methylation of these sequences may have arisen independently of the restriction-modification systems to serve another function. One such function attributed to th *dam* methylase and possibly also to the *dcm* methylase is involvement in mismatch repair and recombination.

The *dam* and *dcm* mutants have no effect on restriction-modification systems, and conversely mutations in restriction-modification genes scarcely affect the level of DNA methylation (Mamelak & Boyer, 1970; Marinus & Morris, 1973; 1974).

6.2.1 Recombination and *dcm*

The *dcm* mutants show no abnormalities with respect to growth, mutagenesis, or survival after ultraviolet irradiation (Szyf *et al.*, 1982; Marinus & Morris, 1973; Bale *et al.*, 1979). Methylcytosines, however, can act as sites of mutational hotspots (see Section 7.2.1).

There are a number of reports that repressor binding is modified when methylcytosine replaces cytosine in mutated operator sequences (Fisher & Caruthers, 1979, and personal communications quoted in Korba & Hays, 1982a). This shows that the base change of a cytosine to methylcytosine *can* affect gene expression, although this process is not normally physiologically significant.

The *arl* mutation in *E. coli* leads to a partial deficiency in methylcytosine content of the *E. coli* DNA, and of phage lambda DNA and plasmids maintained in *arl*− mutants (Korba & Hays, 1982a, 1982b). The methylcytosine content of *arl*− phage lambda DNA is about one-third that of *arl*+ phage DNA, but on infection of *arl*+ bacteria with *arl*− phage, the phage DNA regains its normal level of methylcytosine, showing the change not to be heritable. The *arl*− phage are highly recombinogenic and apparently contain large amounts of hemimethylated DNA. This relationship was confirmed by Korba and Hays (1982a) by constructing hemimethylated DNA by hybridizing DNA from a *dcm*− host (no methylation) with that from a host carrying EcoRII (methylated at CC^A_TGG). This DNA is also highly recombinogenic.

The hyper *rec* phenotype of *arl* mutants distinguishes them from *dcm*− mutants and wild type cells, which either lack methylcytosine or are fully methylated, and implies the presence of a recombination mechanism in *E. coli* that can act on DNA shortly after replication when it is present in a hemimethylated form, as illustrated in Figure 6.2.

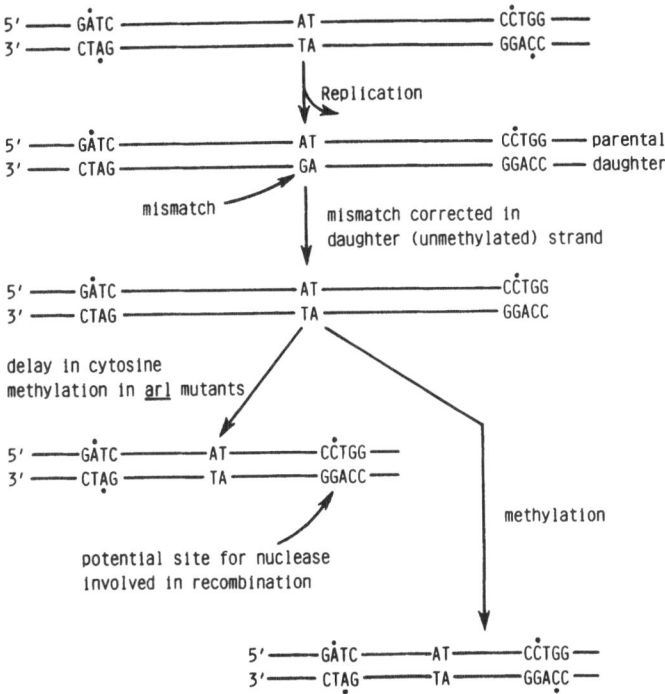

Figure 6.2. Mismatch repair. The delay in methylation of the daughter strand following replication of DNA allows time for mismatches to be corrected. The delay in cytosine methylation is accentuated in *arl* mutants, giving rise to an increased frequency of recombination. • methyl groups.

6.2.2 *dam* Mutants and Mismatch Repair

In contrast to *dcm* mutants, *dam* mutants show high mutation rates, are more sensitive to ultraviolet light and mutagens (Glickman & Radman, 1980; Marinus & Morris, 1974), and are inviable in the double mutants with *recA, recB, recC, lexA,* etc. (Marinus, 1981).

The *dam* methylase methylates every GATC sequence, introducing two methyl groups per site in unmethylated duplex DNA or adding one methyl group per site in the hemimethylated DNA present immediately after replication. Methylation of newly synthesized bacterial DNA takes place very quickly after replication and no more than 3 kbp of DNA are present in the hemimethylated state (Geier & Modrich, 1979; Szyf *et al.*, 1982). Such conclusions are obtained by making use of the unusual specificity of the restriction endonuclease DpnI, which recognizes the sequence GATC only when it is methylated on both strands. Pulse-labeled DNA can be restricted with either DpnI or MboI (which only cuts unmethylated GATC sequences) and separated by gel electrophoresis. Szyf *et al.* (1982) showed that even with the shortest pulses (30 s) the radioactive DNA is sensitive to DpnI and the limits of detection in the assay were such that 20% digestion of such DNA would have been detected. Marinus (1976) came up with a different answer. Looking at the methylation of nascent DNA made during a 2-min pulse with ^{3}H-adenine, he found it to have only 0.96 mole percent methyladenine, as opposed to 1.4 mole percent for total DNA. The difference may result from the fact that DNA methylases may be limiting in some cases. Szyf *et al.* (1984) have shown that both the *dam* and *dcm* methylases are limiting under conditions where extensive replication is occurring, such as following induction of a phage λ lysogen. The methylation changes under these conditions are site specific, implying that the enzymes have a different affinity for the different sites.

Because of the delay in methylation of nascent DNA reported by Marinus (1976) and the consequent presence of hemimethylated DNA near the replication fork, it was suggested that methylation may play a role in mismatch repair (Wagner & Meselson, 1976). The DNA replication system has an extremely high fidelity, based partly on the limited errors introduced by DNA polymerase and partly by the proofreading function present in *E. coli*, which allows removal of a mismatched, newly incorporated base prior to elongation. Errors that escape this initial correction mechanism can be corrected later by excision repair reactions or by reactions involving recombination and catalyzed in *E. coli* by the products of the *rec* genes and the *lex*A gene.

Excision repair can efficiently remove a damaged strand of DNA. For example, following ultraviolet irradiation, the thymine dimers formed produce a distortion in the double helix that is recognized by an endonuclease, resulting in nicks in the damaged strand on the 5' side of the dimer. The 3' OH produced acts as a primer for a strand displacement reaction catalyzed in *E. coli* by DNA polymerase I, shown in Figure 6.3. Excision repair will also replace regions of damage caused by removal of bases by N-glycosylases. For example, when a cytosine residue becomes deaminated to uracil it is recognized as a foreign base and removed by uracil N-glycosylase. A problem arises when the proofreading mechanism fails and a base mismatch is formed in which both of the bases, that on the template strand and the newly incorporated base, are normal. In this case how can a mismatch correction system discriminate between the two normal bases to remove the incorrect one?

Similar but more acute problems arise when a methylcytosine is deaminated to thymine to give a T:G mismatch (see Figure 7.8) and also following homologous recombination if sequence differences occur in the heteroduplex region formed by breakage and rejoining of parental molecules.

It is not clear how these last two problems are resolved, but there is good evidence that the discrimination between an incorrect base in the

Figure 6.3. Excision repair of a thymine dimer.

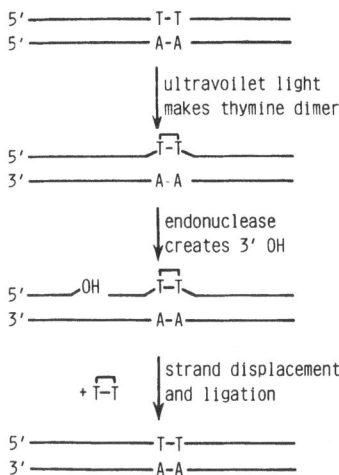

newly synthesized strand and the base in the template strand relies on the transitory undermethylation of the newly synthesized strand (see Figure 6.2) (Radman *et al.,* 1981; Glickman & Radman, 1980; Radman, 1981).

Mismatch correction depends on an additional set of enzymes encoded in the *uvrD, uvrE, mutH, mutL, mutR,* and *mutS* genes, as well as the *dam* methylase (Nevers & Spatz, 1975; Rydberg, 1977; Radman *et al.,* 1978; Marinus, 1981; Lu *et al.,* 1983). In *dam⁻* cells, neither strand of DNA is methylated; the repair system cannot discriminate between parental and daughter strands, and so repair occurs at random. However, in *mut⁻* cells the repair system is absent, and mismatches are not removed, which again leads to a high rate of spontaneous mutagenesis (Glickman & Radman, 1980; McGraw & Marinus, 1980).

By introducing plasmids carrying the *dam* gene into *E. coli,* one can increase the level of *dam* methylase up to 500-fold. Plasmids that increase the amount of enzyme only tenfold have little effect on the mutation frequency, but when very high levels of *dam* methylase are present in cells, the mutation frequency is increased up to 50-fold (Marinus *et al.,* 1984). This, presumably, arises from the increased levels of methylase acting on the nascent, hemimethylated DNA before the mismatch correction enzymes have a chance to work.

The initial experiments supporting a role of *dam* methylation in mismatch repair involved constructing phage lambda heteroduplex DNA differing in adenine methylation and also in genetic markers. When these were transfected into *E. coli* there was a strong bias among the progeny phage for those markers carried on the methylated strand (Radman *et al.,* 1981). The bias was independent of whether the recipient cells were *dam⁺* or *dam⁻* but was abolished in *mut* mutants. The heteroduplexes were made by annealing separated lambda DNA strands isolated from *dam⁺* and *dam⁻*

cells and suffered from the disadvantage that even the former are not completely methylated. Together with incomplete separation of the two lambda DNA strands, this meant that the bias was incomplete. Pukkila *et al.* (1983) increased the efficiency of strand selectivity by methylating lambda DNA to completion *in vitro* with purified *dam* methylase. Heteroduplexes with both chains fully methylated were refractory to repair, but in the heteroduplexes with one completely methylated and one unmethylated strand the frequency of repair on the latter was 100%. Thus it appears that the mismatch repair system does not act on a methylated strand regardless of the state of methylation of the complementary strand.

In experiments in which heteroduplexes were formed between gapped duplexes of M13 RF DNA and single strands carrying markers constructed to have single base changes, there is very poor rescue of the mutagenized marker (Kramer *et al.*, 1982). This is as a result of mismatch repair of the unmethylated single strand and can be corrected by methylation using *dam* methylase.

An *in vitro* assay has been developed using f1 R229 RF DNA containing a base pair mismatch in the single EcoRI site (Lu *et al.*, 1983).

$$5' - G \ A \ A \ T \ T \ C -$$

$$3' - T \ T \ T \ A \ A \ G -$$

In vivo this is repaired to an EcoRI sensitive site if the upper (correct) strand is *dam* methylated and the lower strand is not *dam* methylated, and a similar reaction is catalyzed by *E. coli* cell extracts in the presence of ATP. The mismatch correction activity is dependent on the products of the *mutH*, *mutL*, *mutS*, and *uvrE* genes, and by complementation assay the products of these genes can be isolated for future study.

Mismatches may occur during recombination or, indeed, during excision repair, when both strands are largely methylated. It is possible that the repair patches generated are sufficiently large (and unmethylated) for the bias of the mismatch correction system to generate the correct base pair—otherwise random correction will lead to increased levels of mutagenesis following repair and recombination.

6.2.3 Location of GATC Sites

There has been much speculation as to the significance of the presence of a cluster of *dam* (GATC) sites near the origin of replication of *E. coli* and *S. typhimurium* (Sugimoto *et al.*, 1979; Meijer *et al.*, 1979; Zyskind & Smith, 1980). There are 14 such sequences present in a 296-bp origin fragment, all of which are conserved between the two species, as illustrated in Figure 6.4a. Transformation frequencies of *oriC* plasmids introduced into *dam⁻* cells are greatly reduced compared with those introduced

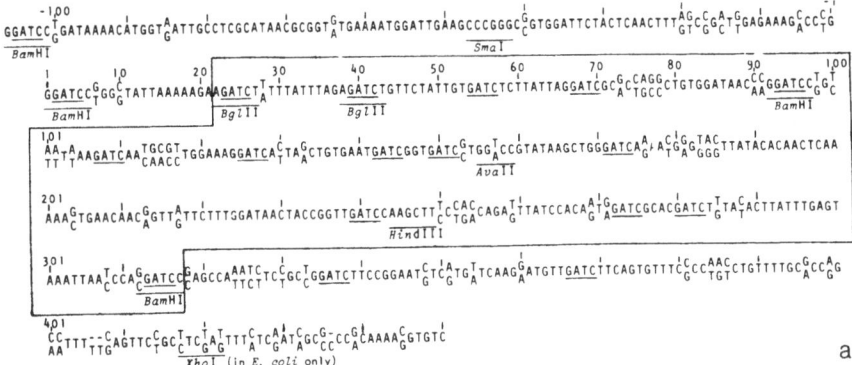

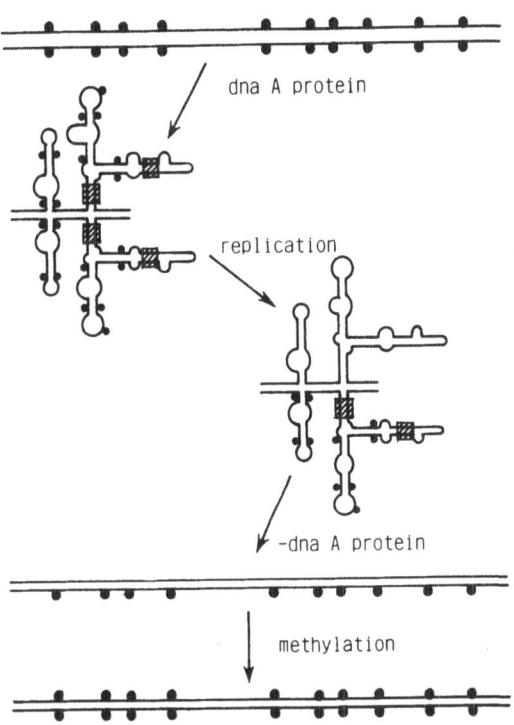

Figure 6.4. (a) The *ori* regions of *S. typhimurium* (top row) and *E. coli* (bottom row) indicating the prevalence of GATC sequences (underlined). Where the sequences are identical a single line is presented. The essential region is shown within the box. (b) A possible mechanism to explain regulation of initiation of replication. Initiation depends on binding of the *dnaA* protein (hatched boxes) to four sites in the *oriC* region which may assume a cruciform configuration. On replication one side of each cruciform remains unmethylated until resolution occurs on complete loss of *dnaA* protein. Filled circles represent methyl groups. (From Zyskind & Smith, 1980.)

into *dam*⁺ cells and this effect is not related to mismatch repair. Even in *dam*⁻ cells a low level of methylation occurs and one particular GATC site in the origin region is always methylated (Smith *et al.*, 1985). This region is also potentially capable of forming an abundance of secondary structures that may have a role in initiation of replication. The finding that the *dam* methylase is capable of enhancing the pairing of complementary single-stranded regions may in some way aid in the formation of such secondary structures (Buryanov *et al.*, 1984). Many of the GATC sites occur in the stem regions of potential cruciform structures and so, on replication would be unmethylated on both strands of the duplex regions as illustrated in Figure 6.4b. This may provide a mechanism for regulation of initiation of replication in that *dnaA* protein may not bind to the unmethylated sites and methylation may be delayed until the cruciform is resolved and the *dam* methylase can act at hemimethylated regions.

In an *in vitro* system, a nonmethylated plasmid carrying the origin region of *E. coli* (*oriC*) is 50 to 80% less efficient in the initiation of DNA synthesis than is the fully methylated plasmid. Initiation at the correct *in vivo* origin is dependent on the presence of the *dna*A protein, but its binding site, though within the *oriC* region, contains no GATC sequences (Hughes *et al.*, 1984; Fuller & Kornberg, 1983).

An alternative explanation for the cluster of *dam* sites at the origin of replication is that they ensure a very low rate of mutation in a region that is essential for survival (Zyskind & Smith, 1980). As we have seen in the previous section, increased mutability is found when GATC sites are methylated before the mismatch correction system gets a chance to act. Unless the *dam* methylase acts in a processive manner (see Chapter 5), it will take very much longer to methylate a string of *dam* sites than a single site, thus ensuring longer exposure to the mismatch correction enzymes and a higher level of conservation of origin sequences.

By cleaving *E. coli* DNA with DpnI, which recognizes fully methylated GATC sites, and scanning a subsequent electrophoretic separation, Szyf *et al.* (1982) concluded that the distribution of GATC sites is random. Small clusters would not be detected, but a regular appearance of sites at specific intervals along the genome would have resulted in a discrete size band, which was not seen. This was contrary to expectations based on an experiment of Gomez-Eichelmann and Lark (1977), who showed that DpnI treatment fails to reduce the size of Okazaki pieces. The latter group concluded that GATC sites must occur at the ends of Okazaki pieces, which they had rendered fully methylated by incubation with AdoMet.

Although methylation of GATC sequences is not essential for survival of *E. coli,* the frequent (if not regular) occurrence of such sequences (methylated or not) may potentiate the initiation of synthesis of RNA along a DNA duplex and may therefore be important for replication and transcription. There is little or no evidence in support of this idea at present, however.

6.3 Modification in Phage mu

The *mom* gene of phage mu specifies an enzyme that modifies adenine residues in the concensus sequence

$$5' - \begin{matrix} C \\ G \end{matrix} A \begin{matrix} G \\ C \end{matrix} N \, Py - 3'$$

(Toussaint, 1976; Kahmann & Kamp, quoted in Kahmann, 1984). The modification, which affects 15% of the DNA adenine residues, results in formation of the unusual adduct α-N-(9β- D-2'-deoxyribofuranosylpurine-6-yl)glycinamide (Swinton *et al.*, 1983), shown in Figure 6.5. This is also known as N^6-(1-acetamido)deoxyadenosine or dA_x' for short. The *mom* enzyme can also act in *trans* modifying not only the mu DNA, but also the host DNA and the DNA of superinfecting phage or plasmids (Toussaint, 1976). This renders the DNA insensitive to several restriction systems *in vivo,* and the modified DNA is resistant to a large number of type II restriction enzymes *in vitro.*

The *mom* gene is only transcribed during lytic development of the phage and, for expression, it requires that the host be dam^+. Thus in dam^- *E. coli* the phage mu DNA lacks the unusual modification showing that it is dispensable for growth (Toussaint, 1976).

In addition to *dam*, a second *trans*-acting protein is required for *mom* expression. This is the product of the phage C gene, which interacts with the *mom* promotor. As this gene is expressed only following induction of DNA synthesis, this explains why *mom* expression is linked to lytic development. It is possible that the C gene product is a phage-coded RNA polymerase or acts to modify the host enzyme (Plasterk *et al.*, 1984). In artificial constructs in which the *mom* promoter is replaced by, for instance, the P_L promoter of phage lambda, expression no longer depends on the phage C gene product. Deletion of a short region upstream from the *mom* gene removes the requirement for *dam* methylation and allows ex-

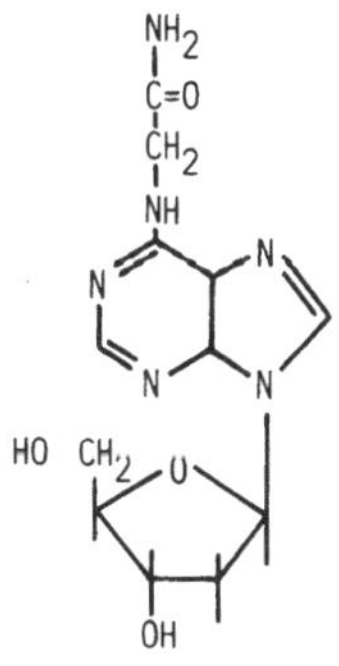

Figure 6.5. N^6-(1-acetamido)-deoxyadenosine.

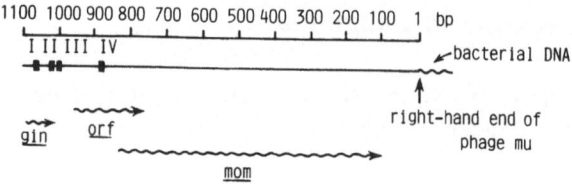

Figure 6.6. The location of four (I–IV) GATC sequences [■] in the 5' flanking region of the *mom* gene at the right-hand end of phage mu.

pression in *dam⁻* cells (Plasterk *et al.*, 1983; Kahman, 1983; Hattman *et al.*, 1983).

This short region contains the sequence (C)GATCG repeated three times within a 34-bp region (I, II, and III), and a fourth GATC sequence (IV) lies 134 bp to the right (Adley & Bukhari, 1984), as is clear from Figure 6.6. Notice that the palindromic repeats contain the GATC recognition site for the *dam* methylase, and it is postulated that the secondary structure of the promotor may depend on the state of methylation of these sequences (Plasterk *et al.*, 1983, 1984), shown in Figure 6.7. Removal of the first of these repeats allows expression in *dam⁻* cells as do small deletions in the region around base 1040 (Figure 6.6). The requirement for *trans*-activation by phage C gene product remains (Kahmann, 1983) showing that dependence on *dam* methylation requires a different region from that involved in *trans*-activation (the promoter). This seems to rule out the location of the promoter between repeats II and III as proposed by Plasterk *et al.* (1983, 1984), and Adley and Bukhari (1984) suggest that the promoter close to the GATC repeats I, II, and III (Figure 6.6) is of a

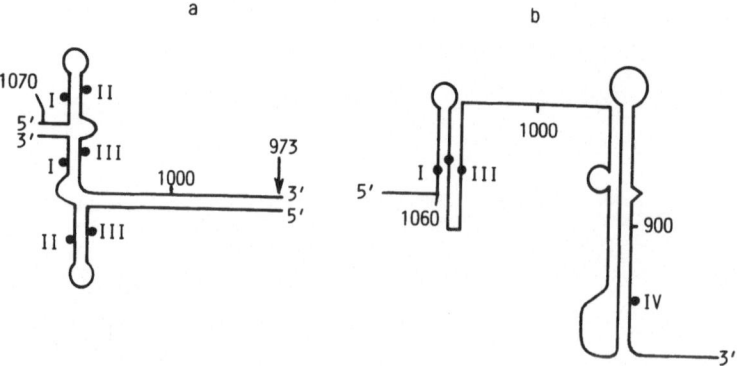

Figure 6.7. Two of the possible secondary structures for the *mom* promoter region as proposed by (a) Plasterk *et al.* (1984) and (b) Kahmann (1983). The numbering is from the right-hand end of phage mu DNA as indicated in Figure 6.6. Dots indicate sites of *dam* methylation.

regulatory nature. When it is active, it occludes promotion from the primary *mom* promotor found about 200 bp downstream, just after GATC IV. Methylation would therefore inhibit transcription from the upstream regulatory promoter, thereby allowing *mom* transcription to be initiated from the primary promoter.

Here we have a clear example of an effect of DNA methylation on secondary structure and on transcription. Why such a control system has developed is unclear but the effect is that the *mom* modification is only present when *dam* methylase is active, *i.e.*, when the mismatch correction system is operational. Perhaps this plays a facilitative role in phage mu replication.

7

Methylation in Higher Eukaryotes

7.1 The Sequence Methylated

For the most part the only methylated base (other than thymine) found in the DNA of higher eukaryotes is 5-methylcytosine (see Chapter 1). As early as 1955 Sinsheimer concluded that this is present in the dinucleotide CG. He isolated dinucleotides resulting from DNase I digestion of calf thymus DNA and removed terminal phosphates with prostatic phosphomonesterase. The dinucleotides were fractionated by ion exchange chromatography and split by venom phosphodiesterase to give a 5′ mononucleotide and a mononucleoside that were separated and analyzed, as shown in Figure 7.1.

The only dinucleotide that was found to contain methylcytosine at the 5′ end was mCG. However, the relative amounts of the various dinucleotides found are not representative of the base composition of the DNA, and in particular the dinucleotide AC was absent. These anomalous findings are a result of the specificity of action of pancreatic DNase I, which acts preferentially on the Pu-Py bond (Young & Sinsheimer, 1965), and therefore, although the evidence for the presence of the dinucleotide mCG is sound, it cannot be excluded that other mC-containing dinucleotides may occur. Although they did not determine the location of the methylcytosine in the dinucleotide, Grippo *et al.* (1968), using a similar approach, showed it to occur predominantly in CG.

Doskocil and Sorm (1962) hydrolyzed DNA with dilute acid, which leads to the removal of the purines and the release of pyrimidine nucleotides in the form p-(Py-p)n. Terminal phosphates were again removed with prostatic phosphomonoesterase, and the pyrimidine runs hydrolyzed

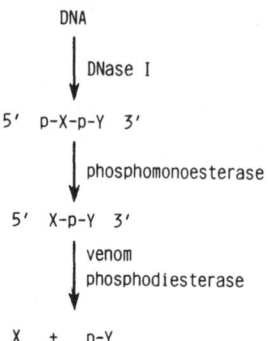

Figure 7.1. Identification of methylated dinucleotides.

with calf spleen phosphodiesterase, which yields 3′ monophosphates. Those nucleotides occurring at the 3′ end of a run of pyrimidines were released as nucleosides.

$$Py\text{-}p\text{-}Py\text{-}p\text{-}Py$$
$$\downarrow$$
$$Py\text{-}p + Py\text{-}p + Py$$

In DNA from calf thymus, rat spleen, and mouse liver, methylcytosine was recovered only in the unphosphorylated form, indicating its occurrence in the dinucleotides mCG and/or mCA. The nature of the purine could not be determined by this method but, in light of Sinsheimer's (1955) results it has been assumed to be only guanine.

Browne and Burdon (1977) obtained pyrimidine runs from a variety of DNA's and separated them by homochromatography. By using DNA prelabeled in methylcytosine they were able to show that the distribution of methylcytosine among the pyrimidine runs was quite distinct from that of cytosine, but it was essentially the same in the DNA of normal and transformed cells from mouse, hamster, human, and *Xenopus*. This finding is consistent with methylcytosine occurring in the same sequence throughout the DNA's tested.

Somewhat contradictory evidence was reported by Sneider (1972), who found substantial amounts of methylcytosine within and at the 5′ end of runs of pyrimidines from the DNA of Novikoff rat hepatoma cells. Thus, about equal amounts of mCT and TmC were found. More recently Kaput and Sneider (1979) have obtained further evidence from studies with restriction enzymes that significant amounts of mCC occur in Novikoff rat hepatoma cells.

Restriction enzyme studies are based largely on the use of the two isoschizomers HpaII and MspI. Both enzymes cleave the sequence CCGG but HpaII is inhibited when the internal cytosine is methylated and MspI when the 5′ cytosine is methylated. When high molecular weight mouse liver DNA is cleaved with MspI, the size of the resultant DNA fragments is close to that predicted from base composition and nearest neighbor analysis (Singer *et al.*, 1979a) indicating that the sequence mCCGG occurs rarely. However, HpaII cuts the DNA into fragments much larger than those found with MspI, and it can be calculated that 77% of the MspI sites are methylated at the internal cytosine, *i.e.*, the methyl-cytosine occurs in the dinucleotide mCG. It was thought that evidence for methylation in mCC had been found when particular sequences in human globin genes were discovered to be resistant to MspI (van der Ploeg *et al.*, 1980). These experiments were done by cleaving nuclear DNA with the restriction enzymes and separating the fragments by gel electrophoresis. The fragments were then transferred to nitrocellulose paper by Southern blotting and probed with ^{32}P-labeled cloned globin DNA (see Section 8.3). Recent experiments (Busslinger *et al.*, 1983a; Keshet & Cedar, 1983), however, suggest that this result is not due to methylation of the 5' cytosine, but to a reduced ability of MspI to cleave certain sequences such as

$$5'—CmCG\ \ GCC—3'$$
$$3'—G\ \ GCmCGG—5'$$

where the methyl group is again present in the dinucleotide CG.

   The presence of methylcytosine can be detected when DNA is sequenced by the Maxam-Gilbert technique (Maxam & Gilbert, 1977; 1980). The hydrazine reagent in dilute buffer that reacts with DNA at pyrimidine residues fails to act at methylcytosines or thymines in high salt so a band is missing on the sequencing gels; as indicated in Figure 7.2 (see Appendix A7). When DNA isolated directly from the cell has been sequenced, methylcytosine has been found only in the sequence CG (cloned DNA loses its methyl groups during growth of the vector in *E. coli*). However, if cytosines in other sequences are methylated, but only occasionally, these may escape detection. Unfortunately the ability to obtain sufficient cellular DNA for direct sequencing is limited to satellite DNA or riboso-mal or histone DNA, which can be easily isolated in pure form by re-peated fractionation on gradients of cesium chloride. Nonetheless to-gether with the previous evidence, this result points strongly to the occurrence of methylcytosine only in the sequence mCG.

   Another approach that has been used by Gruenbaum *et al.* (1981b) is to nick DNA with DNase I or by sonication and then carry out a variation of the nearest neighbor analysis, adding a single [α^{32}P]-labeled dNTP with *E. coli* DNA polymerase I. The labeled DNA is then hydrolyzed with micro-coccal nuclease and spleen phosphodiesterase to the 3' nucleotides, which are separated by two-dimensional thin layer chromatography as shown in Figures 7.3 and 7.4.

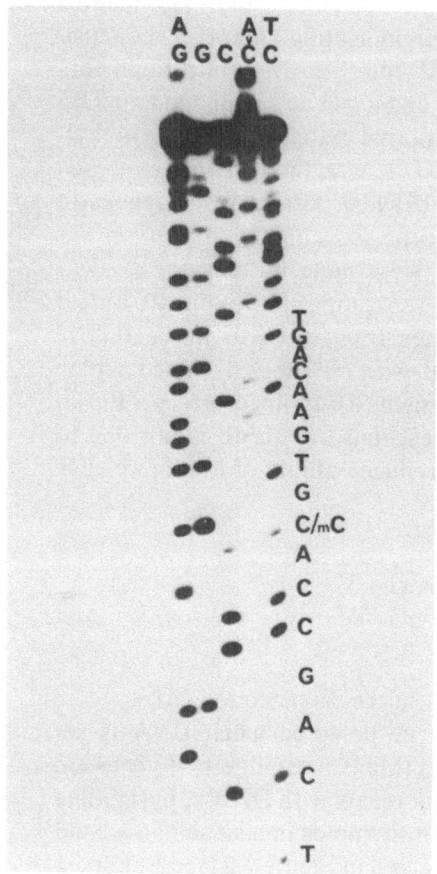

Figure 7.2. Detection of methylcytosine by Maxam-Gilbert sequencing. An uncloned fragment of bovine satellite DNA clearly indicates a band of reduced intensity corresponding to the presence of mCG. (From Pech *et al.*, 1979. Copyright M.I.T.)

Although Gruenbaum *et al.* used only the one [α^{32}P]-labeled dNTP and so could not comment on the occurrence of mCA, mCT, or mCC dinucleotides, they concluded that 80% of CG dinucleotides were methylated in calf thymus DNA, but only about 40% were methylated in rabbit liver, and less than 30% in sea urchin sperm.

By using MspI to cut the DNA, Cedar *et al.* (1979) were able to look at the level of methylation of the internal cytosines in all CCGG sequences. They removed the 3′ phosphates remaining after restriction enzyme treatment and labeled the internal cytosine using polynucleotide kinase and [γ^{32}P]ATP. The DNA was then reisolated and hydrolyzed to the 5′ mononucleotides using pancreatic DNase I and venom phosphodiesterase, as Figure 7.5 illustrates.

The mononucleotides were fractionated by two-dimensional thin layer chromatography, and the ratios of counts in the dCMP and methyl dCMP

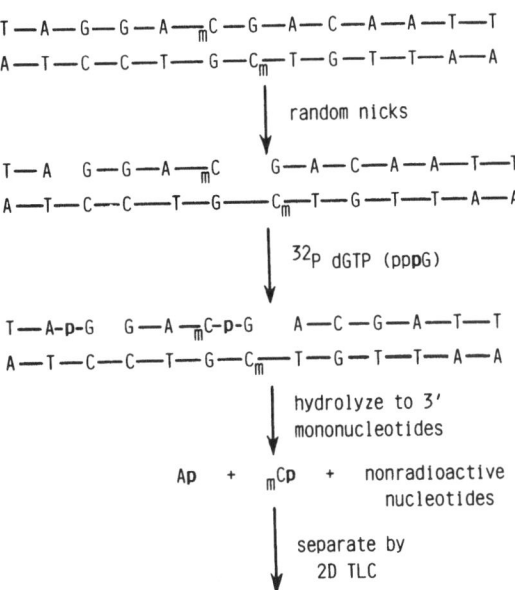

Figure 7.3. Modified nearest neighbor analysis using a single α^{32}P-labeled deoxyribonucleoside triphosphate.

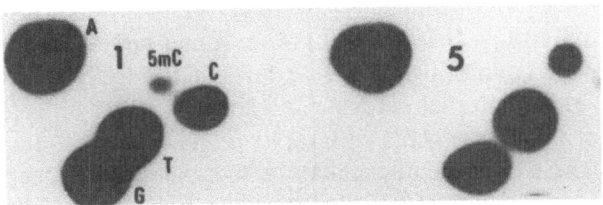

Figure 7.4. Two-dimensional thin layer chromatographic separation of deoxyribonucleotides. DNA was labeled with α^{32}PdGTP and hydrolyzed to the 3' mononucleotides. (1) Mouse mitochondrial DNA. (5) Mouse nuclear DNA. (From Pollack *et al.*, 1984.)

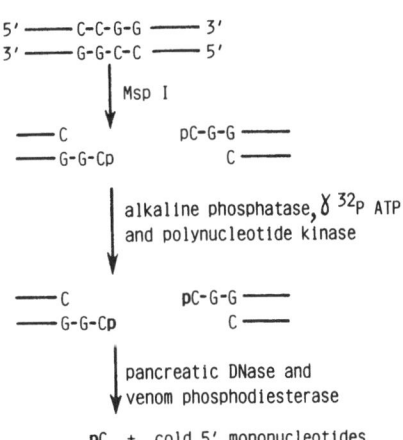

Figure 7.5. 5' end labelling of MspI-cleaved DNA gives the proportion of these sites methylated at the internal cytosine.

spots were compared. With calf thymus DNA the results show that about 90% of the internal cytosines in the sequence CCGG are methylated. This approach also shows that the sequence must be methylated on both strands. With DNA from *Drosophila* less than 0.1% of these cytosines are methylated, and in DNA from sea urchin sperm and rabbit liver intermediate values (15 and 50%, respectively) are found, reminiscent of the results of the modified nearest neighbor analysis referred to above (Gruenbaum *et al.*, 1981b).

Sneider (1980) used a similar approach to look at the level of methylation of the 5' cytosine in the CCGG sequence. After cleavage of DNA with HpaII or MspI, he transferred a radioactive phosphate group to the 5' cytosine by means of the Klenow fragment of DNA polymerase I and [α^{32}P]dCTP. The DNA was then hydrolyzed to 3' mononucleotides, which were separated by two-dimensional thin layer chromatography, as illustrated in Figure 7.6.

With Novikoff rat liver DNA cleaved with MspI no radioactivity was found in 5mdCMP confirming that the action of the restriction enzyme is blocked by methylation of the 5' cytosine, if any occurs in that position. What was surprising was that almost as much radioactivity was found in dCMP from DNA restricted with HpaII as with MspI. This shows that about 90% of CCGG sequences are totally unmethylated in this tissue. Those methyl groups that do occur were found largely in the sequence mCCGG. With bovine liver DNA about half the CCGG sequences are methylated at the internal cytosine and about 40% are unmethylated.

Except for the experiments with rat liver DNA carried out in Sneider's laboratory, there is little evidence for methylation at sequences other than CG, but the percentage of CG dinucleotides methylated is very variable

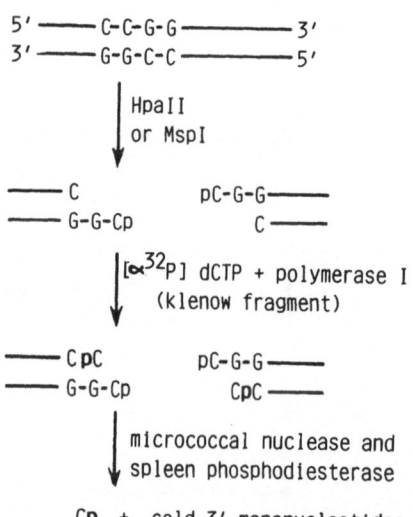

Figure 7.6. 3' end labeling of MspI- or HpaII-cleaved DNA gives the proportion of these sites methylated at the 5' cytosine.

even among vertebrates and in many invertebrates much less methylation is found. The results with the rat DNA should be followed up, as they provide evidence for an unexpected variation which, if confirmed, may necessitate a reappraisal of our ideas about localization of methylcytosine.

In the meantime we must assume that most, if not all, methylcytosine occurs in the self-complementary sequence

$$5' \ mCG \quad 3'$$
$$3' \quad GCm \ 5'$$

As, in practice, not all CG dinucleotides are methylated, the cause and effects of the variation in the level of methylation between various types of cellular DNA's has given rise to much exciting speculation with regard to function (see Chapters 8 to 12).

7.2 Deamination of Methylcytosine

Methylcytosine and cystosine undergo deamination *in vitro* especially in the presence of hot acid (Sneider, 1973; Ford *et al.*, 1980) and alkali (Wang *et al.*, 1982). As these conditions are often used to analyze the base composition of DNA, such methods may somewhat underestimate the methylcytosine content and overestimate the thymine content, as indicated in Figure 7.7.

Deamination of cytosine and methylcytosine *in vivo* would lead respectively to the production of uracil and thymine. Uracil is an abnormal base in DNA and, as such, is recognized by a specific uracil N-glycosylase that initiates a sequence of reactions leading to the replacement of the uracil by the original cytosine (Gates & Linn, 1977). It is very important that this repair sequence functions efficiently so that large numbers of base substitutions do not build up in the DNA. It is perhaps to make this possible that DNA does contain thymine rather than uracil (Lindahl *et al.*, 1977). Primeval DNA molecules containing uracil instead of thymine would not have been able to reverse the effects of cytosine deamination, and such DNA would have tended over evolutionary time to a composition of uracil and adenine only.

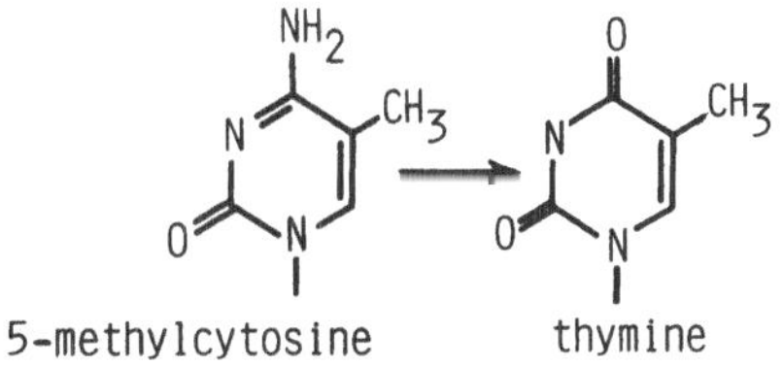

Figure 7.7. The deamination reaction.   5-methylcytosine   thymine

The deamination of methylcytosine to thymine does not lead to the production of an abnormal DNA base, and hence the resulting thymine is not recognized by a repair system. Scarano (Scarano, 1969; Tosi *et al.,* 1972) proposed that such a deamination reaction occurring *in vivo* could be a signal for differentiation during early development. Thus a subgroup of genes might be characterized by a specific array of bases near their promoter, which would be a binding site for a specific DNA-binding protein. Only when the binding protein was bound would the genes be expressed. Binding could be sequentially activated and inhibited by enzymic methylation and deamination, shown in Figure 7.8. Although DNA methylation does occur, no evidence has been forthcoming for a DNA methylcytosine aminohydrolase, and as far as is known deamination occurs only by nonenzymic processes. However Scarano's hypothesis is not too unlike several recent suggestions for the possible role of methylation in the control of transcription and differentiation (see Chapter 8). If deamination of methylcytosine does occur by enzymic or nonenzymic mechanisms, there will be a minor fraction of DNA thymine that is synthesized by a route involving S-adenosylmethionine rather than folic acid enzymes. Scarano *et al.* (1967) produced evidence for such *minor thymine* as support for his model, but such results are bedeviled by the possibility of *in vitro* deamination of methylcytosine that may occur during analysis (Sneider, 1973).

7.2.1   Evidence for Deamination *in Vivo*

Evidence for deamination *in vivo* of 5-methylcytosine has been found in prokaryotes where some spontaneous base substitution hotspots have

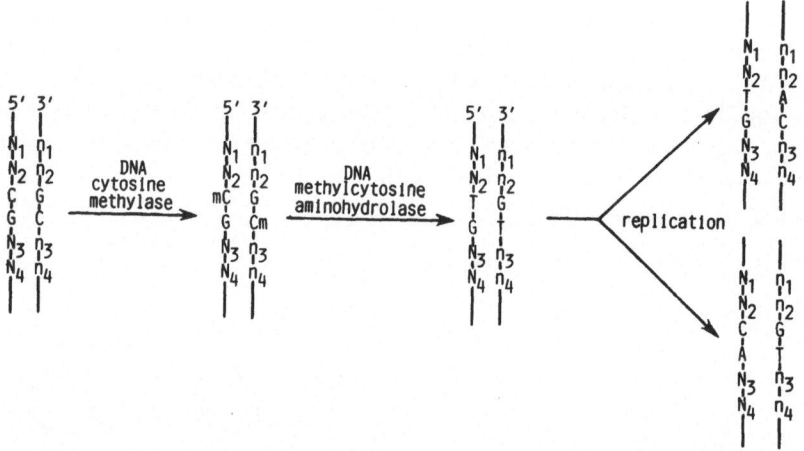

Figure 7.8. Scarano's hypothesis for the chemical basis of differentiation.

been defined as arising by enzymic methylation of cytosine followed by deamination, presumably nonenzymatically, to thymine (Coulondre *et al.*, 1978). Nonsense mutations can arise in the *lac* I gene of *E. coli* by single base substitutions in one of over 80 positions, and Coulondre *et al.* analyzed 298 spontaneous, nonsense mutations. These were not found to be evenly distributed. At two sites the frequency was ten times the average frequency at the remaining sites. Both hotspots occur at the second cytosine in the sequence CCAGG, which is the position methylated by the *dcm* cytosine DNA methylase of *E. coli* K12. These hotspots are not seen in *E. coli* B, which lacks the cytosine methylase.

These deaminations result in a mismatched base pair that would be subject to repair. If the mismatch repair system acts preferentially on the unmethylated strand (see Section 6.3), then it will replace the thymine with cytosine, thereby restoring the status quo. However, if replication intervenes before repair or if the mismatch repair system is insufficiently biased as to the strand on which it will act when only one methylcytosine is missing, then the hotspots will occur.

Similarly, regions showing a high frequency of polymorphism in human DNA have been located at CG dinucleotides. Nine out of ten polymorphic sites investigated by Barker *et al.* (1984) were found at MspI (CCGG) or TaqI (TCGA) sites. These might mutate following methylation and deamination, and this is supportive evidence that this process does occur in eukaryotes. Further evidence, which is discussed below, arises from the observation that in vertebrate DNA in which a high proportion of CpG dinucleotides are methylated, the frequency of these dinucleotides is considerably less than what would be predicted from the base comparison.

7.2.2   Consequences of Methylcytosine Deamination

Most of the DNA in prokaryotes codes for proteins, and hence the process of methylation and deamination must be seriously disadvantageous. These disadvantages may be outweighed on some occasions if the presence of cytosine methylases allows the host bacterium to resist invasion by bacteriophage DNA. However, the number of methylcytosines will be kept to a minimum in order to keep the number of mutations low. In higher eukaryotes where a large fraction of the genomic DNA does not code for proteins, the cell may be able to withstand the pressures posed by higher levels of methylcytosine with the potential for deamination.

In 1972, Comings pointed out that the AT-rich nature of heterochromatic (noncoding?) DNA may be explained by the deamination of methylcytosine. Bird (1980) has shown that eukaryotes may be divided into three groups. The arthropods have little or no cytosine methylation, and the dinucleotide CG occurs at the expected frequency in their DNA. In other invertebrates, for example the sea urchin (Bird *et al.*, 1979), there are methylated and unmethylated compartments, and about 35% of HpaII sites (CCGG) are methylated overall. This intermediate extent of methylation is associated with an occurrence of the dinucleotide CG at only 65% of the frequency expected from overall base composition. Vertebrates form the extreme where about 70% of HpaII sites are methylated but the dinucleotide CG occurs at only 20 to 30% of expected frequency, shown in Figure 7.9. An implication of these findings is that as the extent of methylation has increased it has been accompanied by an increase in deamination so that CG has been converted via mCG to TG and its complement CA. Bird points out that the reduction in frequency of occurrence of the dinucleotide CG is indeed often accompanied by corresponding increases in the dinucleotides TG and CA (Setlow, 1976; Russell *et al.*, 1976). Such a decrease in the CG dinucleotide accompanied by an increase in TG and CA would result in a gradual enrichment of the genome in adenine and thymine and depletion in guanine and cytosine, which might partially explain the imbalance of the bases found in vertebrate DNA.

7.2.3   Protection Against Deamination

This inverse relationship however, does not always hold true (McClelland & Ivarie, 1982; Smith *et al.*, 1983), and the idea of passive loss of methylated CG dinucleotides may be a little too simple. In particular, vertebrate coding regions are in general not low in G+C (and in some cases are enriched in these bases) (Fraser *et al.*, 1975; McClelland & Ivarie, 1982). It is possible that, although passive deamination may be the major influence in noncoding regions, in coding regions there may be active selection for regions where deamination has not occurred. The reason for this selection

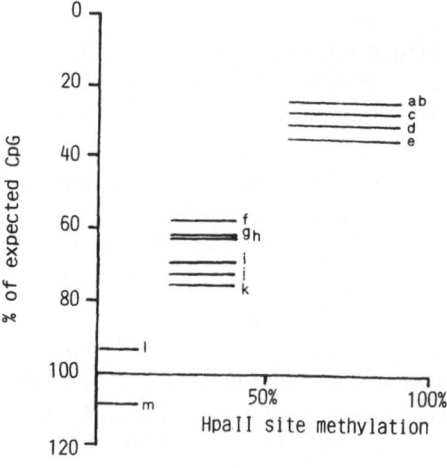

Figure 7.9. Correlation between CpG deficiency (expressed as a percentage of expected frequency) and the level of HpaII site methylation in various animal DNA's: (a) man; (b) chick; (c) mouse; (d) rabbit; (e) hamster; (f) starfish; (g and h) sea urchins; (i) sea anemone; (j) sea squirt; (k) sea cucumber; (l) fruit fly; and (m) bee. (From Bird *et al.*, 1979. Copyright M.I.T.)

is not altogether obvious as is exemplified by a comparison of the coding regions for α and β globin. These two genes are closely related in evolutionary terms, yet the α globin gene is much richer in G+C than the β globin gene, and it has the number of CG dinucleotides expected from a consideration of its base composition. In the coding region these are 25 CGs. In 20 of these cytosine is in the third codon position. That is, when deamination of a methylcytosine in the DNA would have no effect on the amino acid composition of the protein (Salser, 1977; Smith et al., 1983; Lennon & Fraser, 1983). It has also been suggested that selection may be based on differential codon usage, and that globin genes expressed in the embryo may be transcribed in a limiting aminoacyl-tRNA mixture, which demands a biased codon usage for efficient transcription. This idea is not supported by the finding that the gamma globin gene is neither G+C rich nor CG rich (Shen et al., 1981), yet is active in cells also expressing α globin. It is important to realize the constraints placed upon coding for amino acids when the G+C content of the DNA rises much above 50%. Under these conditions there will be selection against codons where the first two bases are G or C, but the base in the wobble position is immaterial. Alpha globin coding regions may have a high proportion of cytosines simply because they have not been selected against. An alternative viewpoint, as suggested by Salser (1977), is that the transcribed RNA may assume a precise secondary structure that would reduce the number of allowed mutations and may have led to the preservation of some genes with an undiminished level of CG dinucleotides.

It is intriguing that it is in G+C-rich DNA that we find these higher levels of CG dinucleotides. In such DNA the chances are high that the CG dinucleotide will occur in the sequence GCG. Indeed this is the case for 42 of the 104 CG dinucleotides in the human enkaphalin sequence (Noda et al., 1982) in the EMBL data bank. Deamination of these 42 GmCG sequences would produce GT dinucleotides that may help to generate spurious splice sites (Breathnach et al., 1978).

The most likely possibility, however, is that it is the increased stability of the double helix in G+C-rich DNA that physically mitigates against deamination of methylcytosine. Chemical deamination of cytosine residues has been used to determine secondary structures in RNA molecules as the reaction occurs significantly more slowly in double-stranded regions than in single-stranded regions (Goddard & Schulman, 1972; Shapiro et al., 1973), and an analysis of the neighbors of the CG dinucleotides found in vertebrate DNA indicates that these regions are G+C rich and would be expected to be considerably more stable than average (Adams & Eason, 1984).

To some extent it is misleading to talk of genes which are A+T rich or G+C rich. Careful examination of the composition of genes in the EMBL DNA data bank reveals wide fluctuations in G+C content over relatively short distances. For example, the human gamma globin gene region, which has an average base composition of 41.4% G+C, shows regions of 100 adjacent bases with 25% G+C and others of 60% G+C. See Table 7.1 and Figure 7.10 for details. This variation is by no means unexpected as a sequence of bases generated to be random except for the constraint that the overall composition is 40% G+C shows regions of 100 bases in which the composition varies between 24% G+C and 60% G+C (Adams & Eason, 1984).

An inspection of Figure 3.3 shows that certain vertebrate gene regions have an exceptionally high G+C content and these genes are not at all deficient in the dinucleotide CG. Here it is not just a small region that has been protected from loss of the CG dinucleotide. Rather the whole gene region may fluctuate around the very high G+C-rich composition. Some of these regions code for ribosomal RNA; for instance the *Xenopus* rDNA region has a composition of 74.4% G+C. Here it is the transcribed regions that are most G+C-rich and the nontranscribed part of mouse rDNA has only 49.8% G+C. Other G+C-rich (and CG-rich) regions are protein coding, but are interspersed by introns of similar composition, *e.g.*, the human enkaphalin gene and the human α globin gene.

Table 7.1. Variation in G+C Content of Some Vertebrate Gene Regions and Random Sequences[a]

	Gene	Average composition (% G+C)	Length scanned (bp)	Min. % G+C	Max. % G+C
1	Chick albumin	39.1	2368	23	51
2	Human β globin	40.7	2052	19	61
3	Human γ globin	41.4	11376	25	60
4	*X. borealis* 5S RNA	49.3	761	33	65
5	Mouse rRNA	52.3	960	37	70
6	Mouse immunoglobulin	54.2	1674	42	64
7	Chick histone H2A	59.0	843	30	80
8	Human enkaphalin	66.7	1014	51	78
9	Human α globin	67.0	1138	52	82
10	Rat rRNA	72.5	2282	49	90
11	*X. laevis* rRNA	74.4	1326	47	98
	Random	30	10^4	15	49
		40	10^4	24	60
		50	10^4	33	65
		60	10^4	41	74
		70	10^4	46	86
		80	10^4	65	93

[a] The sequences from the EMBL nucleotide sequence data library were scanned for composition in consecutive blocks of 100 bases.

Figure 7.10. Variation in C+C content along a series of gene regions. The maximum and minimum values obtained for 100 contiguous nucleotides are plotted against the overall G+C content. The squares are figures from random sequences of defined G+C content. The circles are for (1) chick albumin; (2) human β globin; (3) human γ globin; (4) *X. borealis* 5S RNA; (5) mouse rRNA; (6) mouse immunoglobulin; (7) chick histone H2A; (8) human enkaphalin; (9) human α globin; (10) rat rRNA; (11) *X. laevis* rRNA. The lines are the best straight-line fits for the results from the random sequences. (From Adams & Eason, 1984.)

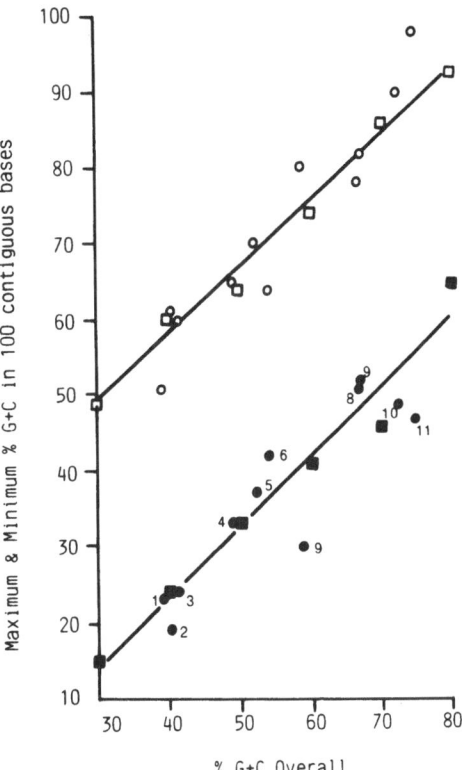

Maximum & Minimum % G+C in 100 contiguous bases

% G+C Overall

A closer inspection reveals that even in these gene regions with a high overall G+C content there are regions of average G+C content. Within each short region the relationship still holds that sequences with a G+C content of 70 to 80% are not deficient in the CG dinucleotide, whereas those with a G+C content of 40% and below are very dramatically deficient. It seems very likely that this mosaic pattern is brought about by the physical constraints placed on deamination by the increased stability of the G+C-rich double helix (Adams & Eason, 1984). In support of this proposal is the finding that only two of the three positions in which methylcytosine occurs in the *E. coli lac* I gene are mutational hotspots (Coulondre *et al.*, 1978). As explained above, those mutants are believed to occur by deamination of methylcytosine, and in the third site where mutation occurs at reduced frequency the methylcytosine occurs in the very GC-rich sequence GGGCmCAGGC.

The finding of a methylated and an unmethylated compartment in the DNA of invertebrates other than arthropods (Bird *et al.*, 1979; Bird & Taggart, 1980) has now been extended to vertebrates (Cooper *et al.*, 1983), where about 2% of the genome consists of DNA unmethylated for stretches of several hundred base pairs (see Section 7.3). This small fraction of DNA appears to be unusually rich in CG dinucleotides and may have been maintained in an unmethylated state to preserve a sequence with an essential function. The location of this fraction in the genome is of considerable interest, especially in the light of the finding by McClelland and Ivarie (1982) that around 15 mammalian genes the CG dinucleotides are very asymmetrically distributed, being enriched in the 5' flanking regions and depleted in the 3' flanking regions. The first two exons of the major histocompatibility complex genes are also enriched in CG dinucleotides (Tykocinski & Max, 1984; Max, 1984), but, as with the 5' regions of certain genes, these are also G+C-rich regions, and so that prevalence of CG dinucleotides may simply reflect the decreased susceptibility of these regions to deamination. The alternative viewpoint that these regions are never methylated is of course also tenable and is supported by the finding that in the DNA from sperm and all other tissues studied, the 5' regions of mouse dihydofolate reductase, hamster adenine phosphoribosyl transferase, and chick α 2-collagen genes are undermethylated (Stein et al., 1983; McKeon et al., 1982). It is not clear, however, why these regions are so rich in G+C bases, because an unmethylated CG dinucleotide would remain equally resistant to deamination whether it was in a G+C-rich region or not.

7.2.4   Methylation and Deamination in Viral DNA

The small animal DNA tumor viruses, such as SV40 and polyoma, resemble their hosts in being deficient in CG dinucleotides, while the larger DNA viruses such as herpes simplex have the expected amount of CG dinucleotides. In neither case is the DNA of the virion methylated, and hence the deficiency of CG dinucleotides in SV40 and polyoma cannot have come about by deamination of methylcytosine during the normal lytic cycle of these viruses. This may point to a recent origin of such viruses from cellular DNA sequences that were already deficient in CG dinucleotides. Alternatively, there may be a frequent interchange of genetic information between sequences of papova virus DNA replicating freely in the cell nucleus and integrated copies of all or part of that DNA. Cells may be exposed to the virus but may not become lytically infected. Instead the viral DNA may be integrated into the host chromosome and inactivated. Such integration occurs during cell transformation, but in this case the DNA is only partly inactivated. The inactivated DNA becomes methylated (see Chapter 8) and so, with time, would become subject to a change in base composition brought about by deamination of methylcytosine to thymine. Reactivation of such sequences would produce the viral DNA deficient in CG dinucleotides that we see today. It is noteworthy that of the 27 pairs of CG dinucleotides present in the SV40 chromosome today, 18 are in the region of the chromosome where initiation of replication and transcription occurs, *i.e.*, in the region where any change in base sequence brought about by deamination may have lethal consequences. It has been shown that while methylation of the cytosines in these sequences does not affect viral gene expression or replication, base changes in these areas do have drastic effects (see Section 8.4).

7.2.5 Summary

To summarize, we are left with the general hypothesis that as the eukaryotes have evolved, so the proportion of the genome that is methylated has increased. It is proposed that because of concomitant deamination this leads to a loss of the CG dinucleotide. However, in G+C-rich regions deamination may not occur, and in other specific regions this loss may have been selected against in one of two ways. If a deleterious mutation occurs the ensuing cell will not survive, and this will lead to selection of cells retaining specific methylated CG dinucleotides. DNA rich in mCG may be actively maintained in cells by gene conversion, serving to correct mutations occurring by deamination. Such a checking and correction mechanism is possible where multiple copies of genes occur, as with the ribosomal RNA genes and various satellite DNA's. Alternatively, mutations by deamination may be prevented by interfering with methylation at certain sensitive sites. Interference may be brought about by the presence of specific binding proteins attached to those sensitive sites at times in development when *de novo* methylases are active (see Section 8.2). This could be the origin of the unmethylated 2% of the vertebrate genome, which may represent a short control sequence upstream of most genes. It follows that methylation of these sequences may well block transcription and hence must be prevented. (It is important to note that this is not a mechanism of physiological control, but a reaction that must at all times be prevented to maintain viability.)

7.3 Location of Methylcytosine in Different Compartments

Within a given cell the level of methylcytosine in DNA isolated from different compartments is different. Compartment is a vague term, but DNA can be fractionated in various ways, *e.g.*, into various frequency classes or into transcribed and nontranscribed regions. The level of methylation of the DNA in the various fractions can be compared.

As methylcytosine in higher eukaryotes occurs almost exclusively in the dinucleotide CG there are two related questions. What is the distribution of CG dinucleotides, and in each fraction what proportion of CG dinucleotides are methylated? Often the answer to the first question is not known, and hence misleading interpretations can easily be put on results, which are usually in the form of the percentage of cytosines or total bases methylated in a given fraction.

In 1979 Bird *et al.* showed, on the basis of sensitivity to the mCpG-sensitive restriction endonucleases HpaII, HhaI, and AvaI, that sea urchin DNA could be divided into two compartments. One compartment is highly methylated and resistant to these enzymes and so remains high molecular weight. The other compartment contains few if any methyl groups and is degraded by these enzymes to low molecular weight fragments, shown in Figure 7.11. The resistant fraction that makes up about 40% of the genome is not devoid of MspI sites, confirming that it is methylation of this fraction that is conferring resistance. This resistant compartment has every CCGG sequence methylated at the internal cytosine in stretches of DNA as long as 50 kbp. All the transcribed sequences are located in the unmethylated compartment, and there is no redistribution of sequences during development of the sea urchin.

Similar methylated (m+) and unmethylated (m−) compartments are found in a wide variety of organisms, *e.g.*, many invertebrates (Bird & Taggart, 1980) and slime molds (Whittaker *et al.*, 1981). On the other hand, using the same criteria the insect genome appears largely unmethylated (Adams *et al.*, 1979b; Rae & Steele, 1979), and the vertebrate genome is largely methylated. However, by end labeling the fragments produced by a restriction enzyme digestion, Cooper *et al.* (1983) were able to show that even the vertebrate genome has an unmethylated compartment amounting to 1%–2% of the genome. 80% of the unmethylated compartment of vertebrate DNA is composed of unique sequences of DNA which are not deficient in CpG dinucleotides. Rather, these are

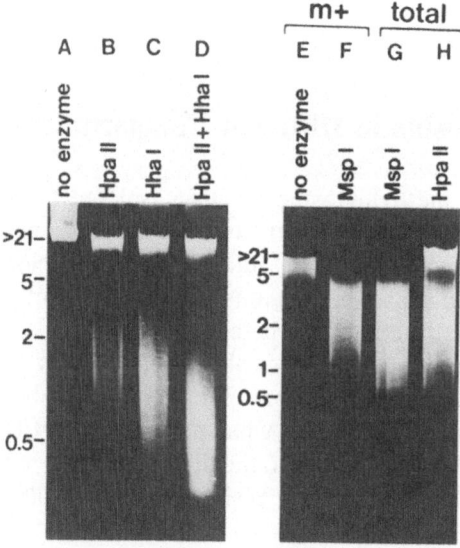

Figure 7.11. Methylated and unmethylated compartments in sea urchin DNA treated with the indicated restriction enzyme and separated by electrophoresis on 1% (left-hand panel) or 2% (right-hand panel) agarose. E and F refer to DNA from the isolated methylated compartment. (From Bird *et al.*, 1979. Copyright M.I.T.)

"islands" of DNA of average size 1 kbp, within which nonmethylated HpaII and HhaI sites are highly concentrated (Bird *et al.*, 1985). Moreover, in the methylated compartment in vertebrates, far from 100% of the CpGs are methylated. By comparing the electrophoretic profiles of MspI and HpaII digests of mouse liver DNA, Singer *et al.* (1979a) concluded that 23% of CCGG sequences were unmethylated. They scanned the photograph of ethidium stained gels, shown in Figure 7.12, and determined the axis that bisected each peak into equal areas. Using a set of standards of known size, researchers could determine the $\overline{Lw}$ (the weight average length of DNA) for the products of a given restriction enzyme digest. For mouse liver DNA this was 4.5 ± 0.5 kbp for MspI and 17.8 ± 0.3 kbp for HpaII. The value predicted on base composition and nearest neighbor frequency information is 3.8 kbp, which agrees quite closely with the MspI result. In DNA of 50 kbp, MspI must cut 21.2 times to give fragments of 17.8 kbp, hence 77% of the total MspI sites are resistant to HpaII, *i.e.*, are methylated.

Careful consideration of the digestion patterns of rodent cell DNA led Kunnath and Locker (1982b) and Naveh-Many and Cedar (1982) to conclude that the unmethylated CCGG sites are not evenly distributed. The discontinuous curves they obtained allowed these workers to conclude that a few percent of the DNA was 30 to 40% methylated; about 70% of the DNA was 70 to 80% methylated; and the remainder was 95 to 100% methylated. (They did not detect any totally unmethylated DNA.) It is not possible at this time to say what these compartments represent but it is highly probable that the undermethylated compartments represent regions of the genome which are being transcribed and which contain clusters of unmethylated CG dinucleotides (see Chapter 8).

7.3.1   Methylation of Nucleosomal Core and Linker DNA

In 1977 Razin and Cedar reported that the percentage of cytosines methylated in DNA isolated from chick nuclei or chromatin treated wih micrococcal nuclease was greater than that of total DNA. The 30 to 70% increase was found in chromatin from a series of chick cells and also from calf thymus. These results were obtained using mass spectrometry of derivatives of purified DNA and were confirmed by Solage and Cedar (1978) using mouse L-cells labeled with [^{3}H-methyl]methionine. In particular, the DNA solubilized at early times of digestion (nucleosome linker DNA) is almost free of 5-methylcytosine.

It is unlikely that this finding is a result of an increased content in the resistant fraction of methyl-rich satellite DNA (see below), because such sequences are not preferentially sensitive or resistant to micrococcal nuclease (Lacy & Axel, 1975). Renaturation analysis of total and mirococcal resistant DNA show the latter to be somewhat depleted in highly reiterated DNA sequences.

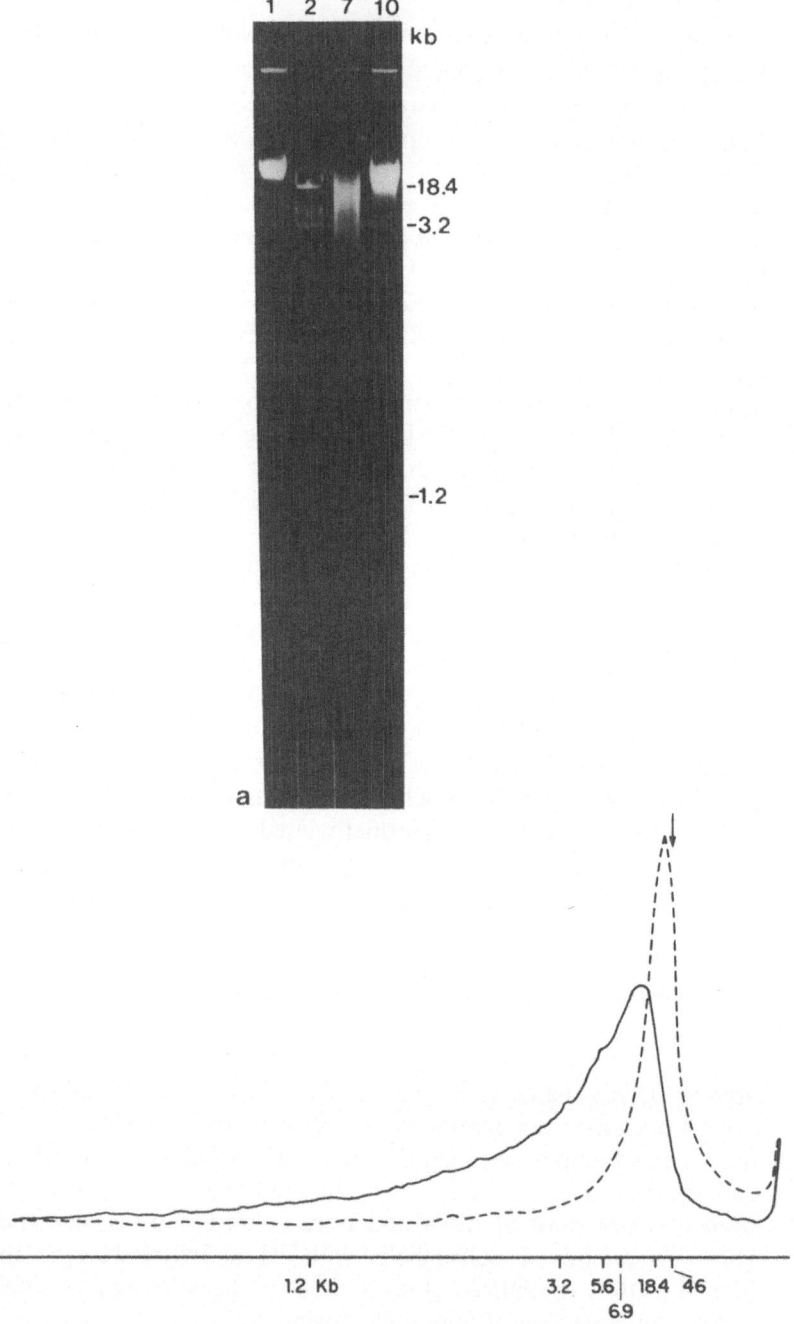

Figure 7.12. HpaII- or MspI-digested DNA from mouse liver separated by electrophoresis. (a) Channel 1: Uncut DNA. Channel 7: DNA cut with MspI. Channel 10: DNA cut with HpaII. Channel 2: Markers. (b) Densitometer tracing of Channel 7 (solid line) and Channel 10 (dotted line). The arrow indicates the position of uncut DNA. (From Singer *et al.*, 1979a.) Copyright 1979 by the AAAS.

Table 7.2. Methylation of Nucleosome Core DNA

	Core DNA	Total DNA
G+C content (mol %)	45.3	41.6
CpG content $\dfrac{(CpG \times 100)}{(NpG)}$	4.23 (81.3)[a]	3.37 (83.8)[a]
Methyl cytosine content $\dfrac{(mC \times 100)}{(C+mC)}$	4.79	3.69

[a] In parenthesis is the % deficiency in CpG content over that expected on a random distribution of bases.

Using less sensitive methods, Adams *et al.* (1977) were unable to find a consistent increase in the methylcytosine content of nucleosome core DNA, but with the advent of high performance liquid chromatography they have confirmed the result of Razin and Cedar (Adams *et al.*, 1984; Davis *et al.*, unpublished data). As shown in Table 7.2 the increased methylcytosine content of DNA in micrococcal nuclease resistant chromatin is associated with an increased G+C and CpG content of resistant DNA and is about what would be expected on this basis.

It seems probable that nucleosomes located on G+C-rich DNA produce a more stable configuration than those located on A+T-rich DNA, and hence there is an increased probability of finding nucleosome core regions in G+C-, and hence, methylcytosine-rich DNA. Alternatively, DNA in linker regions of chromatin may have evolved to become A+T rich as a result of stabilization of core regions against deamination of methylcytosine (see Section 7.2.). However, this second explanation is much less attractive, as it requires nucleosomes to have fixed locations on the DNA and to have maintained those positions for many hundreds of thousands of generations.

7.3.2 Methylation of Satellite DNA

There are a number of reports indicating that rapidly reannealing DNA (highly reiterated and foldback) sequences are more highly methylated than moderately repetitive and unique sequences, as shown in Figure 7.13 (Harbers *et al.*, 1975; Cato *et al.*, 1978; Romanov & Vanyushin, 1981; Drahovsky *et al.*, 1979; Schneiderman & Billen, 1973; Gama-Sosa, Wang, Kuo, Gehrke & Ehrlich, 1983).

The HS-β satellite of the kangaroo rat has a ten base pair repeat with the following consensus sequence.

5'ACACAGCGGG
3'TGTGTCGCCC

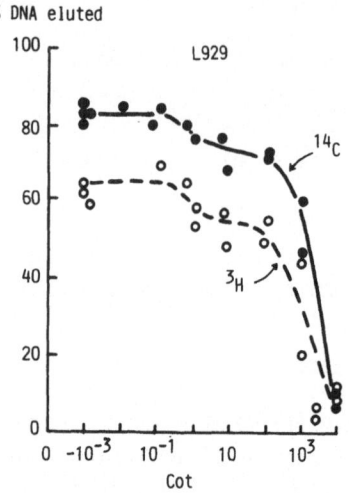

Figure 7.13. Methylation of DNA fractions of differing reiteration frequency. DNA from mouse L929 cells prelabeled with ^{14}C-thymidine and with tritium in methylcytosine as fractionated into different frequency classes using hydroxyapatite chromatography. While 18% of the DNA is rapidly renaturing this contains 35% of the methylcytosine.

(Fry *et al.*, 1973). There is a CG dinucleotide in both strands, and if these are methylated this will give a value for the percent methylation of:

$$\frac{\text{methylcytosine} \times 100}{\text{cytosine} + \text{methylcytosine}} = \frac{2 \times 100}{7} = 28.6\%$$

or

$$\frac{\text{methylcytosine} \times 100}{\text{total base}} = \frac{2 \times 100}{20} = 10 \text{ mol } \%$$

(In practice the composition is only 6.7 mole %, as in more than a third of the repeats the methylcytosine is replaced by guanine or adenine.)

Similarly bovine satellites are believed to have evolved from a dodecamer containing one CG dinucleotide (Taparowsky & Gerbi, 1982), and the mouse satellite, though A+T-rich, is believed to have arisen by amplification of a short sequence containing a CG dinucleotide (Manuelidis, 1981).

It has been pointed out (Adams *et al.*, 1983a) that the presence of mCG dinucleotides in the short sequences that are amplified to form satellite DNA and in interspersed repetitive DNA (*e.g.*, Alu sequences) may be more than an unusual coincidence. It is possible that such sequences have a propensity for amplification and that this is in some way related to one of the functions of methylation in eukaryotes (see Section 7.3.4).

These examples serve to illustrate the point that the high levels of methylcytosine found in satellite DNA are no more than what might be expected from their CG dinucleotide content. Because of this, a large proportion of the methylcytosine of the genome may be in the rapidly reannealing fraction. For example, in the bovine genome, the various satellites make up about 23% of the total DNA, and almost 20% of the cytosines are methylated compared with less than 3% in main band DNA (Macaya *et al.*, 1978; Poschl & Streek, 1980). As the satellites are largely G+C-rich (*i.e.*, have a higher percentage of cytosine than main band DNA), this means as shown in Table 7.3 that in bovine cells more than half of the methylcytosine is in satellite DNA.

There is a certain amount of tissue variation in the figures reported for the level of cytosines methylated in reiterated DNA fractions. It is not clear to what extent this reflects the changes in amount of a repeated DNA fraction in the genome, but this appears to be of minor significance (Cato *et al.*, 1978; Romanov & Vanyushin, 1981; Ehrlich *et al.*, 1982). It certainly does not account for the major reduction in the level of methylation found in satellite DNA from bovine sperm and brain, and human sperm and placenta (Kaput & Sneider, 1979; Sturm and Taylor, 1981; Pages & Roizes, 1982; Sano & Sager, 1982; Gama-Sosa *et al.*, 1983a; 1983c; Adams *et al.*, 1983a). To explain the near zero levels of methylcytosine found for instance in satellite DNA from bovine sperm, we must first determine the period when satellite DNA is unmethylated and study the function of such DNA in order to try to find a role for methylcytosine.

The presence of unmethylated satellite DNA has been studied most carefully for spermatogenesis in the mouse (Ponzetto-Zimmerman & Wolgemuth, 1984; Sanford *et al.*, 1984; Chapman *et al.*, 1984; Feinstein *et al.*, 1985b). In mouse both the major and minor satellites are resistant to digestion with HpaII in somatic cells but are partly undermethylated, not only in mature sperm, but also throughout spermatogenesis at least as early as the premeiotic spermatogonia. The undermethylation is by no means complete in these cells, and in most cases the bulk of the satellite DNA probably has fewer than one in three to five CG dinucleotides unmethylated. These satellite DNA's are also undermethylated in mouse oocytes. Dispersed repetitive and most unique DNA sequences are heavily methylated in sperm, but in oocytes some of the dispersed repetitive DNA is also undermethylated.

In the early embryo, the cells of the primitive ectoderm that give rise to the fetus contain fully methylated satellite DNA's as far as can be as-

Table 7.3. Methylation of Bovine Satellite DNA

	Total DNA	Satellite DNA
% of DNA	100	23
G+C content (mol %)	44	60
cytosines as a % of *total* bases	22	6.9
% methylation $\dfrac{(mC \times 100)}{(C + mC)}$	5.4	13
methylcytosine as a % of total bases	1.2	0.9

sayed, but in the extraembryonic structures (trophectoderm and primitive endoderm) the satellite DNA's are again substantially undermethylated.

It is not clear whether the germ cell satellite DNA becomes undermethylated as a result of suppression of *de novo* methylation that occurs in early mouse embryos (see Section 8.4.6) or by a later, specific demethylation event involved in selective replication of satellite DNA in the absence of methylation.

Satellite DNA (but not interspersed repetitive DNA) is transcriptionally inactive and located in constitutive heterochromatic regions often at or around centromeres and telomeres (Bostock, 1980; Miller *et al.*, 1974). These satellite sequences have evolved from short sequences by amplification associated with frequent point mutations. However, as the multiple copies of a satellite of one species resemble each other more than they resemble the satellite DNA's of related species, it is clear that there must be some mechanism of gene conversion acting to minimize the appearance of heterogeneity.

Centromeres and telomeres are important in chromosome recognition and pairing. In some organisms, satellite DNA's are lost from or reduced in somatic cells, and it has been suggested that satellite DNA plays a role in the pairing and recombination events that occur at meiosis. It is possible that the presence of methylcytosine interferes in some way with this function, and so during gametogenesis the methylation of satellite DNA is actively suppressed. Only later does the level of methylation of satellite DNA increase to prevent unwanted recombination events occurring in somatic cells. This may be the result of the loss of inhibitory proteins, shown in Figure 7.14. This rationalization however, would not explain the lower levels of satellite DNA methylation found in brain cells, nor does it account for the concerted changes in methylation of *interspersed* repeated sequences observed when normal mouse spleen is transformed with the Friend erythroleukemia virus (Smith *et al.*, 1982). It may be pertinent that the DNA found associated with the nuclear matrix is frequently enriched in repeated sequences (Matsumoto, 1981; Razin *et al.*, 1978; Small *et al.*, 1982). It would be interesting to know whether or not this association is dependent on the level of methylation of the satellite DNA or vice versa.

7.3.3 Methylation of Ribosomal DNA Sequences

In both prokaryotes and eukaryotes there are multiple copies of the ribosomal DNA transcription unit (Adams *et al.*, 1981a). Each unit produces a single transcript, which is processed to give the major ribosomal RNA's together with 5.8S ribosomal RNA (in eukaryotes) or 5S and transfer RNA's (in prokaryotes). Figure 7.15 depicts the situation in *Xenopus*.

The ribosomal DNA units are repeated in tandem in eukaryotes but in a dispersed manner in prokaryotes. In some cases the repeating units are

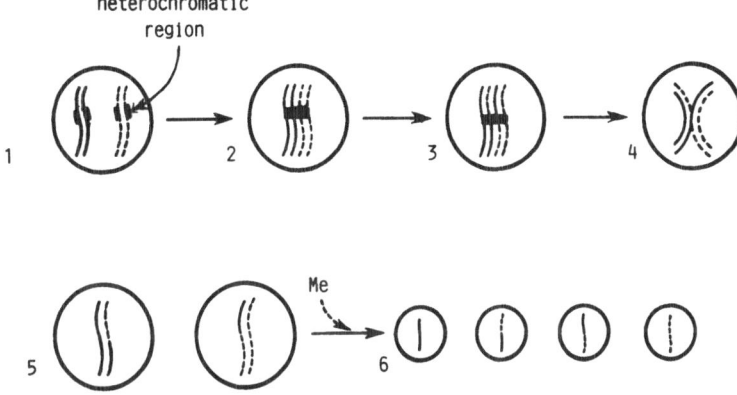

Figure 7.14. Chromosome pairing and recombination, including the changes occurring to the chromosomes during meiosis and a possible role for methylation in the control of recombination. Each line represents a double helix of DNA. (1) *Leptotene:* Following replication in the absence of methylation the two homologous pairs of chromosomes align themselves in the early stages of meiosis. Methylation may be blocked by binding of a protein to specific reiterated sequences. (2) *Xygotene:* Pairing initiates at a specific site to form the synaptonemal complex. (3) *Pachytene:* Recombination nodules indicate the position where crossover events occur. These nodules may represent regions where recombination enzymes recognize unmethylated DNA sequences. (4) *Diplotene and Diakinesis:* The pairs of chromosomes separate, except at the chiasmata where they are held together as a result of recombination. (5) *First Metaphase, Anaphase, and Telophase:* The replicated chromosomes segregate. (6) *Second Meiotic Division Produces Gametes:* Methylation of satellite DNA may take place at this time or it may be delayed, as in Artiodactylae, until after fertilization of the egg.

extrachromosomal (*e.g.*, in *Tetrahymena* and *Physarum:* Gall, 1974), but in higher eukaryotes they are localized at the nucleolus organizer. In *X. laevis* cells there are 400 to 500 copies of the ribosomal DNA repeat unit in the single nucleolus organizer. *Drosophila melanogaster* has two nucleolus organizers, each with 250 copies of the repeat, but in mammals specific nucleolus organizers are seldom visible, and the 200 or so copies

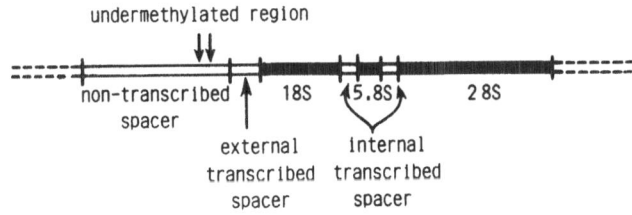

Figure 7.15. One ribosomal DNA repeat unit of *Xenopus laevis*.

of the ribosomal DNA repeat are distributed unevenly over ten or so chromosomes (Miller *et al.*, 1981).

In most invertebrates the ribosomal DNA is not detectably methylated (Bird & Taggart, 1980), whereas in amphibians the chromosomal ribosomal DNA is almost completely methylated (see below).

In mammals most of the ribosomal genes are transcriptionally active and are largely unmethylated (Bird & Taggart, 1980; Bird, Taggart & Gehring, 1981; Kunnath & Locker, 1982a; Miller *et al.*, 1981), as measured using the HpaII/MspI isoschizomeric pair or an antibody specific to methylcytosine. However, in some mammalian cells the ribosomal DNA units are amplified (up to ten-fold in the H4.IIE.C3 rat hepatoma line), and the amplified sequences have a different chromosomal structure indicating lack of transcriptional activity (Miller *et al.*, 1981; Tantravahi *et al.*, 1981; Bird, 1984b). These amplified sequences are enriched in methylcytosine and are interspersed among methyl-poor transcribing ribosomal DNA repeat units to produce extra long differentially staining regions. They are also often present in abnormal chromosomal locations and appear to be unstable and to generate unequal sister chromatid exchanges.

There are at least four different situations, and it is not clear what their interrelationships are:

1. The ribosomal DNA regions are amplified;
2. The amplified regions are not transcribed;
3. The amplified regions are highly methylated; and
4. Recombination events are commonplace.

It is possible that the initial presence of multiple copies of the ribosomal DNA repeat units leads to a high probability of recombination events that can produce amplification. This may be more likely when the gene region is not methylated (see Section 7.3.4). It is important for the cell's well-being that the increased numbers of ribosomal genes are not all transcribed, and this must be controlled in some way. Methylation may be a part of this control, but more likely control occurs as a consequence of the inactivation of the amplified genes. When the gene for dihydrofolate reductase is amplified under selection pressure it also results in abnormal chromosome structures, but it remains transcriptionally active and unmethylated. Even in normal mouse cells there is evidence that the few methylated ribosomal RNA genes are transcriptionally inactive (Bird, 1984b).

The multiple copies of ribosomal DNA present in *Xenopus* somatic cells comprise about 0.2% of the chromosomal DNA (Dawid *et al.*, 1970), and these are extensively methylated. However, methylation is not complete. At one HhaI site (GCGC) in the region coding for 28S RNA about half the repeats are unmethylated and most repeats are cleaved by HpaII (CCGG) and AvaI (CPyCGPuG) in the nontranscribed spacer (Bird & Southern, 1978; Bird, Taggart & Macleod, 1981). This cleavage is a result of the presence of at least one unmethylated site in two clusters of restriction enzyme sites in the 60-base-pair repeated sequence that occurs upstream of the start of transcription. In *X. laevis,* sperm DNA is fully methylated at these sites, but loss of the methyl groups occurs on the first day of embryonic development at about the time when transcription of the ribosomal genes begins (Bird, Taggart & Macleod, 1981). The absence of methyl groups does not appear to be necessary for transcription, but rather may be a consequence of the binding of a protein to a promotor site blocking DNA methylation (Macleod & Bird, 1982, 1983; Pennock & Reeder, 1984). Thus completely methylated (sperm) DNA and unmethylated (cloned) ribosomal DNA are transcribed equally well in the *Xenopus* oocyte system. Moreover in *X. borealis* the undermethylated sites are also present in sperm DNA (Macleod & Bird, 1982). Similarly heavily methylated 5S DNA is transcribed when injected into *Xenopus* oocytes (Brown & Gurdon, 1977), and it is possible that the oocyte has the ability to override any inhibitory effects of methylation. This ability may be essential in order to reactivate many methylated genes during early development of amphibians and may provide a contrasting system to that present in mammals (see Section 8.4.6 for further discussion on this point). During gametogenesis in *Xenopus,* amplification of the already large number of ribosomal DNA repeats occurs in two stages (Bird, 1977). A 10- to 40-fold amplification occurs in both sexes when fewer than 16 primordial germ cells are present (Kalt & Gall, 1974). At meiosis the amplified DNA is lost in the male, but in the female further amplification occurs to give over a million copies of the ribosomal DNA region, which now forms the major fraction of the DNA in the oocyte. This amplified DNA is present in an extrachromosomal form and is not methylated (Dawid *et al.,* 1970; Bird, 1977). It is the amplified oocyte DNA that is transcribed to form the large amount of ribosomal RNA present in oocytes. Transcription of this DNA occurs at an extremely high rate, and the DNA appears free of nucleosomes. It is probable that this very high transcription rate interferes with DNA methylation, or alternatively it may be the extrachromosomal location that does not allow binding of the DNA methylase (*cf.* viral DNA, see Chapter 8).

7.4 Methylation and Recombination in Eukaryotes

A high level of mutation and homologous recombination occurs when cloned (unmethylated) or prokaryotic DNA is introduced by transfection into mammalian cells (Wilkie *et al.,* 1980; Calos *et al.,* 1983; Vincent & Wahl, 1983). Chromosomal rearrangements are associated with activation and demethylation of immunoglobulin genes (Rogers & Wall, 1981; Mather & Perry, 1983; Storb & Arp, 1983), and when aberrant rearrange-

ments occur these may be associated with the activation of cellular oncogenes (see Rabbitts, 1983; Le Beau & Rowley, 1984). Chromosomal rearrangements are also associated with tumors arising as a result of
treatment of Chinese hamster embryo fibroblasts with the demethylating
agent 5-azacytidine (Harrison *et al.*, 1983).

Further evidence for a possible connection between DNA methylation
and recombination comes from studies of sequences involved at or near
recombinational junctions. Slightom *et al.* (1980) have evidence for a
hotspot for recombination between human gamma globin genes in which a
short run of CG dinucleotides occurs in the middle of a longer sequence of
alternating TG dinucleotides. Duplication of the human haptoglobin genes
has generated a CGCG sequence (Maeda *et al.*, 1984), and the cleavage
that brings about the joining of transfected DNA sequences in mouse cells
frequently occurs at a CG dinucleotide (Anderson *et al.*, 1984). Four out
of five immunoglobulin light chain J regions sequenced by Sakano *et al.*
(1979) terminate in a CG dinucleotide. The fifth (which also lacks two
other CG dinucleotides present in the other four J segments) has not been
shown to produce immunoglobulins and may never be linked to a V
region.

These recombination events may be initiated by the action of an endonuclease whose activity is modulated by methylation of DNA. The
enhanced recombination that occurs in *E. coli arl* mutants is attributed to
the presence of undermethylated DNA (Korba & Hays, 1982a) (see Section 6.2.1). Although the initiating endonuclease has not been isolated, the
sequence specificity need not be high. Like the type I endonucleases or
the enzymes involved in mismatch repair, they may act at random on
DNA once it has been recognized as being unmethylated. In higher eukaryotes such nuclease action would initiate recombination between short
homologous regions, a process that occurs to a much greater extent during gametogenesis when reiterated sequences of DNA are undermethylated (see Section 7.3.2). This could lead to the observed high levels of
meiotic crossing over.

As the patches of DNA produced on repair of damaged regions of DNA
tend to be undermethylated (Kastan *et al.*, 1982), and as demethylating
events are reported to increase on aging (Romanov & Vanyushin, 1980),
greater levels of recombination may occur in old, damaged DNA, leading
to heightened levels of chromosomal rearrangements and an increased
likelihood of oncogenic transformation (see Chapter 10). Perhaps it is no
coincidence that tumor cells are often found associated with lower levels
of DNA methylation (Gama-Sosa, Slagel, Trewyn, Okenhandler, Kuo,
Gehrke & Ehrlich; 1983; Feinberg & Vogelstein, 1983a).

8

Methylation and Its Relationship with Transcription

It has been speculated that methylation of DNA may exert a controlling influence over gene expression. This could take the form of switching off genes that are not to be expressed or switching genes on when their transcription is required. Several models have been proposed and these are considered in more detail in Chapter 12. Support for the basic idea came initially when careful analysis showed that the methylcytosine content of DNA varies between tissues as well as during development (the evidence for this is reviewed in Sections 8.1 and 8.2). The results were initially obtained by base composition analysis of isolated DNA, which was very tedious until the advent of HPLC (see Section 2.2 and Appendix). The amount of methylcytosine is usually expressed as moles percent of total bases, or more conveniently as the percentage of cytosines methylated (mC × 100/C + mC) as this takes into account variations in base composition. The other more recent method involves comparing the sizes of fragments found on digestion of DNA with the restriction enzyme HpaII. This enzyme will not cut its recognition sequence (CCGG) if the internal cytosine is methylated, and hence a measure of the proportion of such sequences containing methylcytosine can be obtained, as Figure 8.1 (Singer *et al.*, 1979a). This method is much more of an approximation, and moreover only measures a small fraction of the methyl groups (4–6%), but it compensates for the underrepresentation of the CpG dinucleotide in vertebrate DNA (Section 7.2). It is important that the initial DNA is of a consistently high molecular weight (see Appendix for more details of the method).

As we go through Sections 8.1 and 8.2, the disadvantages of this approach will become obvious, and Section 8.3 deals with correlations found between the methylation and expression of particular genes. These

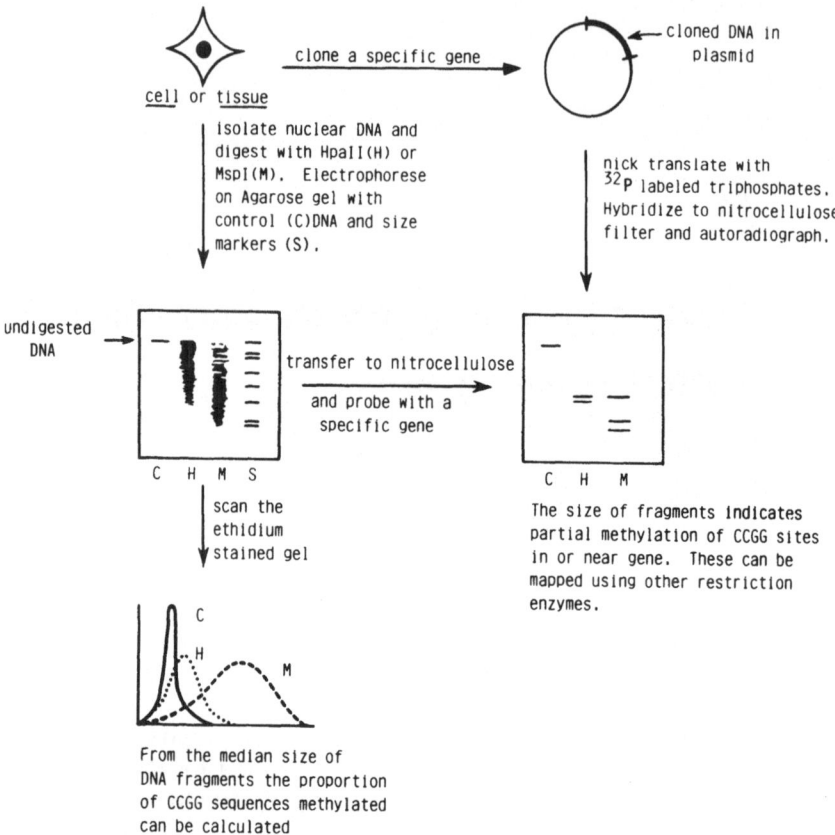

Figure 8.1. Measurement of the proportion of unmethylated CCGG sequences and their locations in specific genes. The left-hand side of the figure illustrates the effect of digestion of vertebrate genomic DNA with either MspI or HpaII (for a photograph of an actual gel see (Figure 7.12). On the right-hand side is shown the type of result obtained when a Southern transfer of the gel is hybridized with a specific gene clone. In the MspI digest, the probe hybridizes to three fragments, but the CCGG sequence joining to two smaller fragments is methylated and hence resistant to HpaII. (Singer, Roberts-Ems & Riggs, 1979.)

experiments are largely based on the use of HpaII and other CG restriction enzymes, *i.e.,* enzymes that have a CG in their recognition site and whose action is blocked by methylation (see Table A.1). Following cleavage of purified nuclear DNA with the restriction enzyme, the digest is fractionated by electrophoresis on an agarose gel. When stained with ethidium bromide, the DNA is found to be smeared throughout the gel. Following transfer to nitrocellulose paper, the position of fragments of a particular gene can be located using a radioactive gene sequence (or part of the gene) as a probe. It is thus essential for this method to have a cloned gene sequence that can be made radioactive for use as a probe (see Figure 8.1).

Results of this type often show a correlation between undermethylation of genes and their transcription. However, in order to show that it is methylation that represses transcription, it is important to introduce into a cell or an *in vitro* transcription system DNA that is either methylated or not, and such experiments are described in Section 8.4.

8.1 Tissue Variation in the Overall Level of DNA Methylation

Table 8.1 lists the proportion of cytosines methylated in DNA from a series of tissues from different mammals. Similar figures are available for the chick, other mammals, frog, and fishes (Vanyushin *et al.*, 1970; Vanyushin, Mazin, Vasiliev & Belozersky, 1973; Kappler, 1971), and numerous authors report figures for a small number of tissues.

Two conclusions can be drawn from these results. First, it is clear (and has been shown to be significant to the 1% level) that not all tissues have the same amount of methylcytosine in their DNA. However, before we rush to try to discover the significance of this finding we should be aware that in different species the pattern differs, as illustrated in Figure 8.2, and

Table 8.1. Tissue Variation in Adult Mammals

Species										
	Methylcytosine content $\left(\frac{mC \times 100}{C + mC}\right)$ in									
	Rat			Mouse		Human	Monkey (4 types)				Bull
Reference Tissue	a	b	d	b	f	c			b		e
Thymus	—	4.43	—	4.68	—	4.79	4.95	5.19	—	—	—
Brain	4.38	4.20	4.37	4.68	3.12	4.69	5.00	5.34	4.32	—	—
Spleen	4.27	4.10	4.29	4.34	3.07	4.45	4.56	4.85	4.71	4.56	6.24
Liver	4.53	4.10	4.28	4.63	2.74	4.21	4.17	4.66	4.71	4.13	6.36
Heart	4.45	3.87	4.12	4.10	2.39	4.16	4.22	4.47	4.76	5.00	—
Kidney	4.08	3.77	4.51	4.39	3.13	—	—	—	—	—	6.00
Lung	4.26	3.92	4.35	4.10	—	4.36	4.22	4.47	4.47	4.37	6.27
Sperm	—	—	—	4.05	—	4.02	—	—	—	—	3.39

[a] Vanyushin *et al.*, 1973a.
[b] Gama-Sosa *et al.*, 1983c.
[c] Ehrlich *et al.*, 1982.
[d] Vanyushin, Mazin, Vasiliev & Belozersky, 1973.
[e] Vanyushin *et al.*, 1970.
[f] Singer *et al.*, 1979b.

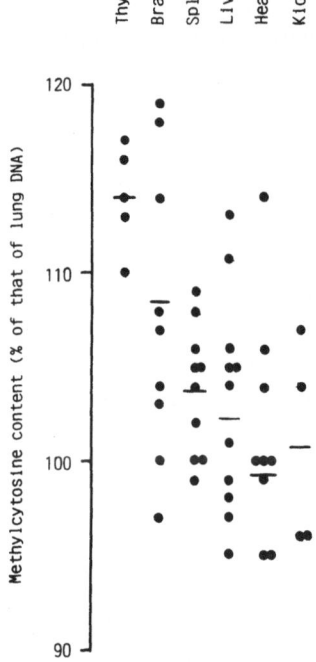

Figure 8.2. Methylcytosine content of the DNA of various tissues from a variety of animals expressed as a percentage of that of lung DNA. The results of Table 8.1 are used.

sometimes the same authors on different occasions obtain differing results for the same species. Although this may be a consequence of using animals of different ages (Vanyushin *et al.*, 1973a), it makes it very difficult to relate a particular value to the function of the tissue. Thus on some occasions brain DNA is the richest in methylcytosine content, having a 19% higher value than lung DNA in the Macaque monkey, whereas it has only 97% as much as lung DNA in the Rhesus monkey (Gama-Sosa *et al.*, 1983c).

It was *partly* on the basis of variations in methylcytosine content of DNA from different tissues that it was proposed that methylation may play a role in the control of gene expression, although further doubt was cast on the importance of tissue variations by reports that methylation patterns can change quite rapidly in cells in culture. Early passage human fibroblasts show a range in the percentage of HpaII sites (CCGG) methylated (59%, 54%, 58.5%, 55%, and 52% in different strains), and this figure was found to rise (in one case) or fall as the culture aged (Reis & Goldstein, 1982a). Probing individual clones with cDNA probes to six different genes revealed striking interclonal heterogeneity, and further heterogeneity was shown to arise *de novo* in pure clones during serial passage (Reis & Goldstein, 1982b). To what extent these changes are a result of chromosomal aberrations occurring in culture is unclear, as the methylation patterns of essential, expressed genes appeared much more constant than those of genes that were not transcribed. Alteration in the chromosome makeup of a cell may explain the differences found in methylcytosine content of normal and virally transformed hamster cells (Cato *et al.*, 1978).

The variation discussed so far is only slight and of doubtful significance. However, Vanyushin's group showed that samples of sperm DNA from bulls, sheep, and pigs have only half to two-thirds as much methylcytosine as DNA from other tissues (Vanyushin *et al.*, 1970). The figure for bull sperm has recently been confirmed in a number of laboratories (Kaput & Sneider, 1979; Sturm & Taylor, 1981; Adams *et al.*, 1983) and has been shown to be caused by the presence of unmethylated satellite DNA which, when fully methylated in somatic cells, contributes a major fraction of the cells' methylcytosine (see Section 7.3.2). Sperm DNA from other mammals is only slightly undermethylated (Ehrlich *et al.*, 1982; Adams *et al.*, 1983a). This may reflect differences in spermatogenesis in different mammals, different functions for satellite DNA, or a different proportion of satellite DNA in the genome.

8.1.1 Induced Changes in Tissue Methylcytosine Content

In order to accentuate any possible relationship between rate of cell division or gene expression and DNA methylation, comparison has been made between (a) normal and regenerating liver, (b) hormone-stimulated and unstimulated tissue, (c) normal and tumor cells, (d) differentiated and undifferentiated cells, and (e) young, adult, and aging animals.

Removal of two-thirds of the liver of the rat stimulates many of the remaining cells to enter the cell cycle, make DNA, and divide 20 to 30 hours later. Although it has been reported (Lapeyre & Becker, 1979) that this procedure is associated with a 12% decrease in the methylcytosine content of the liver DNA, Gama-Sosa *et al.* (1983c) could find no change following partial hepatectomy, but they did show a marked (16%) increase in the first two weeks after birth in the methylcytosine content of liver DNA. The proportion of cytosines methylated in fetal liver DNA was fairly constant at about 3.54%, but rose to 4.10% at two weeks postnatal, whereafter it again remained fairly constant. According to these authors the DNA of other tissues showed little change in development, though Vanyushin *et al.* (1973a) showed a 14 to 17% fall in methylcytosine content of rat brain and heart DNA between 1 and 12 months after birth. The changes in liver DNA methylation may be associated with the failure of many liver cells to divide following DNA replication, giving rise to multinucleate cells. Alternatively, they may be associated with the loss of hematopoietic cells and the changes in function of hepatocytes that occur in the liver around the time of birth. This could be consistent with the transient hydrocortisone-induced increase in methylcytosine content of liver DNA reported by Vanyushin *et al.* (1973a), although other workers have failed to repeat this finding (Baur *et al.*, 1978a; Lapeyre *et al.*, 1980). Moreover, the increase reported occurred largely in the rapidly renaturing fraction of DNA rather than in unique sequences (Romanov & Vanyushin, 1980). Turkington and Spielvogel (1971) found no change in methylation of mammary tissue following hormonal stimulation.

Age-related changes in DNA methylcytosine content have also been reported for salmon, where methylcytosine levels fall to half in various tissues during spawning (Berdyshev *et al.*, 1967), and cows, where thymus and heart DNA lose up to 27% of their methylcytosine content in the 12 years after birth (Romanov & Vanyushin, 1980). In this case, the loss is again from moderately and highly repeated sequences.

Strains of erythroleukemic cells can be obtained by transformation of mouse hematopoeitic tissue (spleen) with certain viruses. Friend cells (transformed by the Friend virus) do not make hemoglobin unless induced by one of a number of chemicals of which dimethylsulfoxide (DMSO) is the most frequently used. A few days following treatment with inducer, cell division ceases, and the cells become red with the hemoglobin they have made. By direct assay uninduced cells have about the same level (3.7%) of cytosines methylated as do other mouse tissues, and satellite DNA is highly methylated (Christman, 1984). There are, in fact, a number of different clones of Friend cells, and analysis of HpaII digests suggests that at least certain clones have only 50% as much methylcytosine as their parent spleen cells (Smith *et al.*, 1982). In particular, an interspersed, repeated sequence has lost its methylation in most locations, implying that the transformation has brought about a concerted demethylation at certain sequences. Following induction of Friend cells, the overall level of methylcytosine remains constant as does the size distribution of DNA following digestion with HpaII. However, DNA isolated from induced cells is a four- to fivefold better acceptor of methyl groups when used as a substrate in a DNA methyltransferase assay (Christman *et al.*, 1977). To what extent this small change, revealed only by a very sensitive assay, is in any way a cause of differentiation is open to question in light of the much larger changes that are observed as random events occurring in cultured cells.

Finally in this section it must be pointed out that although some authors report a reduced level of DNA methylation in nongrowing cells in culture (Rubery & Newton, 1973), other authors found no difference in the level of methylcytosine between log phase and stationary phase cells (Adams *et al.*, 1984; Ehrlich *et al.*, 1982), though lower levels of methylation may arise by growth in limiting methionine (see Section 2.2).

8.2 Changes in DNA Methylcytosine Content During Embryonic Development

Little or no change in the methylcytosine content of DNA isolated from early developmental stages of sea urchins or frogs (*Xenopus*) has been detected (Pollock *et al.*, 1978; Baur *et al.*, 1978a; Baur & Kroger, 1976), although that DNA made in the first round of DNA synthesis following fertilization is not fully methylated until several further rounds of cell division have been completed (Adams, 1973). The early development of mammals differs considerably from that of coelenterates and amphibians in that, in the latter two, the oocyte and egg are huge in comparison to somatic cells, and little increase in mass of the embryo occurs until early development is complete. Indeed, in *Xenopus*, transcription is not initiated until midblastula when 12 cleavage cycles have occurred (Newport & Kirschner, 1982). Furthermore, in amphibians and some less developed organisms, *e.g., Planaria*, "limb" regeneration occurs by redifferentiation of cells that have already differentiated (Slack, 1980), implying that no permanent change to the cellular DNA can have occurred.

However, in mammals the egg is relatively small and most cells remain totipotent only up to the 8-to-16-cell stage (Rossant, 1977). At this time the blastomeres show increased adhesiveness and establish intercellular connections. Division of the 8-cell stage blastomeres generates two types of cells, one of which will form the trophectoderm and the other the inner cell mass. A monoclonal antibody raised against a 124,000 molecular weight cell surface protein interferes with this differentiation (Shirayoshi *et al.*, 1983). The cells of the inner cell mass do not become committed until they are exposed to the blastocoele (Pedersen & Spindle, 1980), but it is these that will eventually form the embryo while the trophectoderm cells form the supporting tissues.

Teratocarcinoma or embryonal carcinoma (EC) cells arise in tumors of the ovary or testis or in early embryos transplanted into extrauterine positions. They are undifferentiated cells carrying surface antigens similar to those present on normal 8-cell embryos (Shirayoshi *et al.*, 1983). When incorporated into an early mammalian embryo they can give rise to normal chimeric adults in which descendants of the teratocarcinoma cells are present in many different tissues (Mintz & Illmensee, 1975; Papioannou *et al.*, 1975). Under appropriate culture conditions teratocarcinoma cells will remain undifferentiated *in vitro* and grow rapidly. If they become confluent, they detach from the substratum and begin a differentiation process somewhat akin to early embryonic development (Evans, 1972; Martin & Evans, 1975).

Singer, Roberts-Ems, Luthardt, and Riggs (1979) could find no difference in the proportion of HpaII (CCGG) sites methylated between mouse embryos at the 2- to 4-cell stage, morula and blastula, and adult mouse tissues. Neither could they detect significant changes in methylation during 13 days of differentiation of one teratocarcinoma cell line, though a second line did show a 22% increase in methylation by 13 days when assayed by HPLC. No difference was found in the size of fragments following HpaII digestion. The reproducibility of some of these results means that the small changes reported are probably not significant, and Fabricant et al. (1979) have shown that different mouse teratocarcinoma cell lines can have widely different methylcytosine contents (3.8–5.2% of cytosines methylated). One of these lines showed a 10% decrease in methylation on differentiation. Young and Tilghman (1984) by analyzing individual genes showed that there is a genome-wide loss of methylation in induced, silent, and constitutively expressed genes when mouse F9 teratocarcinoma cells are induced to differentiate. This was confirmed by Razin et al. (1984), who showed a similar demethylation to occur in extraembryonic membranes, the yolk sac, and the placenta. The mouse embryo DNA itself, however, remains highly methylated. In contrast to most genes, the H2K histocompatibility antigen gene is reported to be very poorly methylated in F9 teratocarcinoma cells, but one site in particular becomes methylated when the cells are induced with retinoic acid (Tanaka et al., 1983). However, this site may be exceptional as the other sites conform to the general finding that genes become demethylated on induction of teratocarcinoma cells (Razin et al., 1984).

In rabbit the much larger trophoblast is easier to obtain and represents 95% of the late blastocyst cells. It resembles most adult tissues in that about 48% of HpaII sites are methylated (Manes & Menzel, 1981; Singer, Roberts-Ems, Luthardt & Riggs, 1979). However, in early cleaving embryos and in the embryoblast (the other 5% of cells in the late blastocyst), more than 70% of HpaII sites are methylated. The suggestion is that while it remains totipotent, the xygote remains extensively methylated, but after nine to ten divisions the unipotent trophoblast has lost methylcytosine.

It is very difficult to interpret the results given in Sections 8.1 and 8.2. Clearly differences do occur in the proportion of cytosines methylated, but these probably represent unimportant variations in most cases, though they could build up to have serious consequences (see Chapter 10). If methylation is related to gene expression, then the values recorded represent the sum of the methylcytosine content of some thousands of genes, many of which may be controlled independently. It is not surprising that small and variable changes are seen. What is surprising is that gross changes are seen and that the overall level of methylation of the inner cytosine in the HpaII sequence is about 70% in the mouse and only 50% in the rabbit.

It is partly because of inconsistent results of experiments investigating overall levels of DNA methylation that the methylation of specific cytosines in particular genes in different tissues has been investigated.

8.3   What Really Is the Evidence for an Inverse Correlation Between Methylation and Gene Expression?

Whenever a research group clones a eukaryotic gene it is a simple step to use this as a probe to investigate whether or not the CCGG sequences are methylated in DNA isolated from different tissues and hence determine if any relationship exists between methylation and gene expression. Often such studies are linked with experiments to determine the level of nuclease sensitivity of the chromatin containing the cloned gene; in particular, experiments to locate DNase I hypersensitive sites. Whenever possible, the effect of the demethylating agent, azacytosine, on expression is also investigated. Besides using cloned genes as probes, it is also possible to use viral DNA to investigate the state of integrated viral DNA in transformed cells.

8.3.1   Chromatin Structure and Methylation

It is possible, even without a probe, to demonstrate that active regions of chromatin are undermethylated. DNA in transcriptionally active regions of chromatin is more sensitive to nicking with DNase I than is DNA in inactive regions. Naveh-Many and Cedar (1981) used such nicks as an entry point for *E. coli* DNA polymerase I and nick translated the transcriptionally active DNA in nuclei using $\alpha[^{32}P]dGTP$ alone. The use of the single triphosphate means that the existing DNA strands are not replaced but that guanines are added to the 3' end of the nicks when these are required by the template strand. The original methylcytosines are thereby retained. On subsequent hydrolysis of the DNA to 3' mononucleotides, only 30% of CpG dinucleotides from active regions were methylated, compared with 66% in total chick erythrocyte DNA, shown in Figure 8.3. The selective nicking of active regions of chromatin is dependent on the presence of the nonhistone, high mobility group (HMG) proteins (Weisbrod, 1982). The HMG proteins can be extracted from nuclei with low salt, and nick translation of such extracted nuclei, as expected, fails to show the presence of a fraction of DNA with a low level of methylation.

In a further experiment to look at transcriptionally active DNA, Naveh-Many and Cedar (1981) hybridized nick translated DNA to poly

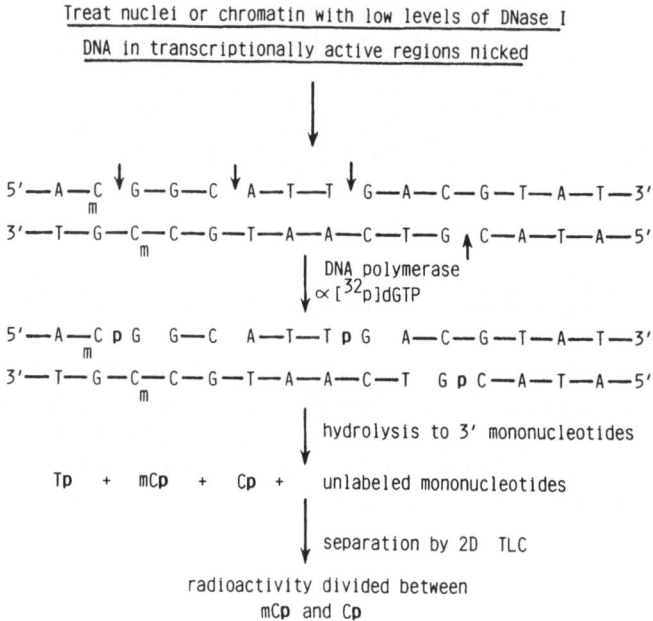

Figure 8.3. DNase I-sensitive chromatin is undermethylaed. Chromatin is nicked with DNase I (in transcriptionally active regions), and α[³²P]dGTP added to the exposed 3′ OH groups. The DNA is hydrolyzed to the 3′ mononucleotides, which on separation indicate the proportion of CpGs methylated in sensitive regions.

A⁺mRNA to isolate the DNA complementary to messenger RNA. For mouse liver, such hybridizing DNA had only 25% of CpGs methylated, as opposed to 65% for total DNA.

Using HPLC, Davis *et al.* (unpublished result) and Davie and Saunders (1981) have analyzed the DNA from transcriptionally active chromatin of mouse L929 cells and calf thymus. This chromatin was isolated on the basis of its solubility in 2 *mM* Mg²⁺ following micrococcal nuclease digestion of nuclei (Sanders, 1978) and found to have 3.24% (L929) or 4.05% (calf thymus) of cytosines methylated compared with 3.96% (L929) or 6.52% (calf thymus) for inactive regions.

As about 10 to 12% of the DNA was found in the transcriptionally active fraction, this shows that 90 to 94% of methylcytosines are in the nontranscribing DNA. These figures are not so marked as those obtained by comparing the proportion of CpGs methylated, and this may reflect the uneven distribution of CpGs.

Using monoclonal antibodies specific for methyl cytosine, Ball *et al.* (1983) have shown that chromatin fractions enriched in HMG proteins are 1.6 to 2.3-fold undermethylated. In contrast, chromatin in which histone

H1 replaces the HMG protein contains 80% or more of the cellular methylcytosine.

8.3.2 Enzyme Induction Associated with Demethylation

Christman *et al.* (1977) showed that treatment of Friend erythroleukemia cells with ethionine induces expression of the globin genes, and the same group (Creusot *et al.,* 1982; Christman, 1984) have more recently shown similar effects with 5-azacytidine and 5-azadeoxycytidine. These drugs are known to lead to synthesis of undermethylated DNA (see Chapter 5.4), but it would be surprising if this alone were the cause of the specific induction produced in these cells. In part it may be the cytotoxic effect of the drugs that leads to induction and concomitant inhibition of cell division.

In a similar manner Jones and Taylor (1980) showed the induction of a differentiated state in cultured mouse embryo cells by a number of cytosine analogues with an altered 5 position, *e.g.,* 5-azacytidine (Figure 5.15). As indicated in Figure 8.4, a small proportion of the treated 10T1/2 cells differentiated into muscle cells, adipocytes, and chondrocytes (Constantinides *et al.,* 1978; Taylor & Jones, 1979; Taylor *et al.,* 1984), and other cell lines have been shown to undergo stable differentiation or viral induction or activation on treatment with low levels of azacytidine (Darmon *et al.,* 1984; Niwa & Sugahara, 1981; Groudine *et al.,* 1981; Reitz *et al.,* 1984). Several rounds of DNA replication following drug treatment are required before one can see the onset of differentiation, by which time the azacytosine has been diluted out, and the DNA has become largely remethylated, though certain particular sites may remain unmethylated. For instance, Darmon *et al.,* (1984), showed induction in mesenchymal cells of only one specific cytokeratin, and Groudine *et al.,* (1981) found that one retroviral gene remains hypomethylated and transcriptionally active following drug treatment.

We shall consider the effect of azacytidine on the expression of specific genes later (sections 8.3.3 and 8.3.4), but it is difficult to understand how such a broad-front effect as the inhibition of methylation throughout the genome could produce expression of specific genes unless they were in a state predisposed to activation. The balance between the undifferentiated and differentiated states may be easily upset, perhaps as a result of a specific demethylation, or perhaps by some other drug-induced action that leads to a reduced level of methylation in specific genes following or associated with changes in gene expression. Whereas activation of a retroviral provirus may require only one demethylating (or similar activating) event and so may occur at lower, less toxic, drug concentrations, the induction of a process of differentiation may require multiple demethylations (or activations) at specific sites, and hence will occur only against a

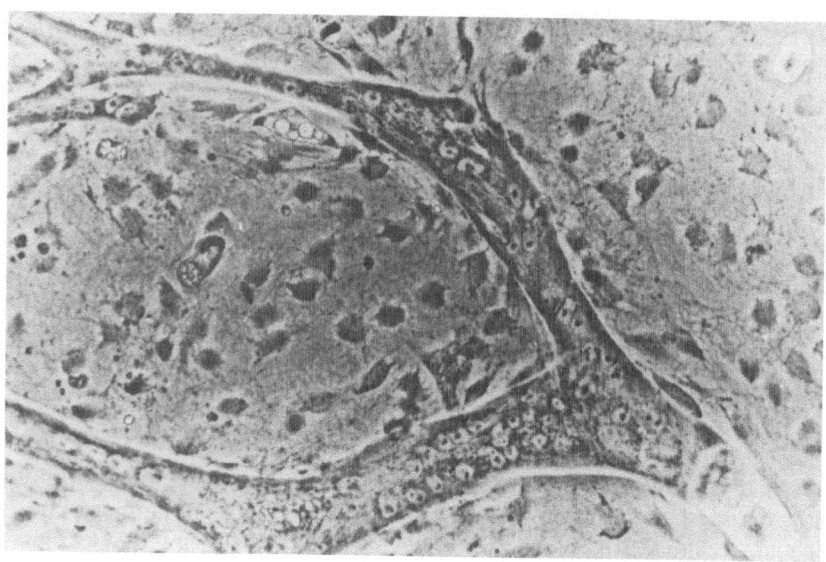

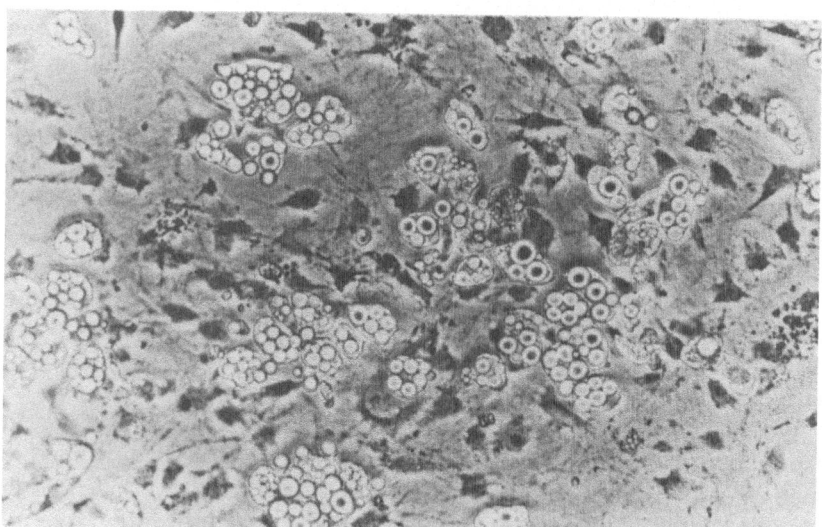

Figure 8.4. Differentiation of cultured mouse embryo cells on exposure to aza-cytidine. The photograph shows adipocytes induced in 10T½ mouse embryo cultures four weeks after treatment with 5-azacytidine (lower) and myotubules induced two weeks after treatment (upper). (From Taylor & Jones, 1979. Copyright M.I.T.)

background of cells dying from the effects of the high doses of azacytidine required.

8.3.3   Genes that Show a Correlation Between Expression and Undermethylation

Table 8.2 lists examples of genes that show a correlation between expression and undermethylation. As pointed out in the introduction to this section, most of these studies rely on the inhibition by methylation of the action of the restriction enzyme HpaII. This enzyme recognizes the sequence CCGG, and hence the results consider only about 4% of the total CpG dinucleotides, but it is reasonable to assume that those recognized are typical of the total CpG population. However, HpaII sites are not very common and will arise on average only about once every 3000 bp in vertebrate DNA, and hence many gene regions will have only one or two sites from which conclusions may be drawn.

With the exception of the DNA of frog virus 3 (FV3) in which 20% or more of the cytosines are methylated (Willis & Granoff, 1980) and Epstein-Barr virus (Diala & Hoffman, 1983), it seems clear that, in general, DNA isolated from intact viruses and free intracellular viral DNA does not contain any methylcytosine. Yet when this DNA is integrated in transformed cells, some if not all of the viral DNA is methylated. The region always methylated is that coding for virion proteins, which are not made in transformed cells (Ford et al., 1980; Desrosiers et al., 1979; Vardimon et al., 1980; Sutter & Doerfler, 1980; Doerfler et al., 1983; Kruczek & Doerfler, 1982; Doerfler, 1984; Miller & Robinson, 1983). A similar correlation exists for retrovirus proviral DNA (Cohen, 1980; Groudine et al., 1981; Stuhlmann et al., 1981; Breznik & Cohen, 1982; Mermod et al., 1983; Hu et al., 1984).

8.3.3a   Adenovirus DNA

No more than traces of methylcytosine (<0.04%) or methyladenine (<0.01%) have been found in adenovirus type 2 virion DNA (Eick et al., 1983), and even these levels may arise by contamination of the viral DNA with tiny amounts of cellular or bacterial DNA. When cellular DNA sequences are encapsidated in virions, they appear to lose their methylcytosines (Deuring & Doerfler, 1983), implying that failure to methylate is in some way tied into the mechanism of replication or location of the extrachromosomal DNA. Free intracellular adenovirus DNA is not significantly methylated in productively infected cells (Vardimon et al., 1980), as measured by sensitivity to HpaII.

Table 8.2. Correlation Between Undermethylation and Gene Expression

A Genes showing good correlation	Reference
Hepatitis B virus	Miller & Robinson, 1983
H. saimiri virus	Desrosiers et al., 1979
Adenovirus	Vardimon et al., 1980; Doerfler, 1984
Mouse mammary tumor virus	Cohen, 1980; Hu et al., 1984
Endogenous chick retrovirus	Groudine et al., 1981
Moloney leukemia virus	Stuhlmann et al., 1981
Rabbit β globin	Shen & Maniatis, 1980
Chicken α and β globin	McGhee & Ginder, 1979
	Weintraub et al., 1981
Ovalbumin	Mandel & Chambon, 1979
Metallothionein	Compere & Palmiter, 1981

B Genes showing imperfect, patchy or no correlation	Reference
Human β globin	van der Ploeg & Flavell, 1980
Immunoglobulins	Rogers & Wall, 1981; Mather & Perry, 1983; Young & Tilghman, 1984
Rat hepatoma protein	Nakhasi et al., 1981
Xenopus tRNA	Talwar et al., 1984
Xenopus rRNA	Bird, Taggart & Gehring, 1981
Human growth hormone Human chorionic somatomammotropin }	Hjelle et al., 1982
Rat prostatic steroid binding protein	White & Parker, 1983
Chick delta crystallin	Bower et al., 1983; Grainger et al., 1983
Rat insulin	Cate et al., 1983
Human insulin	Ullrich et al., 1982
Xenopus vitellogenin	Gerber-Huber et al., 1983
Chick vitellogenin	Wilks et al., 1982; 1984; Meijlink et al., 1983
Collagen	McKeon et al., 1984
Rat α fetoprotein	Kunnath & Locker, 1983; Vedel et al., 1983
Rat albumin	Ott et al., 1982
β casein	Young & Tilghman, 1984
Rat growth hormone	Lan, 1984
Adenine phosphoribosyltransferase Dihydrofolate reductase }	Stein et al., 1983
Histocompatibility antigen H-2K	Tanaka et al., 1983

When adeno 2 or adeno 12 virus transforms hamster cells, the early genes are usually transcribed into mRNA, while late genes are permanently switched off. The methylation state of the CCGG and GCGC sites in three Ad12-transformed hamster cell lines have been precisely measured (Kruczek & Doerfler, 1982). The early genes may produce transcripts of six regions (E_{1a}, E_{1b}, E_{2a}, E_{2b}, E_3, and E_4), but E_{2b} was not expressed in any of these three transformed lines, and E_3 was silent in one transformed line, as shown in Figure 8.5. In all cases where a gene is expressed, those sites in the promoter-leader sequence and at the 5' end of the gene are unmethylated. Sites within the body of the gene may or may not be methylated. Nonexpressed early regions are usually, but not always, entirely methylated, but all late regions are completely methylated. In some adeno 12 virus-induced rat brain tumors some late viral genes are expressed, and, in such cases, these late genes are undermethylated (Sutter & Doerfler, 1980). In adeno 12 virus-induced tumors in hamsters, however, the level of methylation of all viral sequences is low (Kuhlmann & Doerfler, 1982). This anomalous result may reflect a low level of transcription of all viral sequences and possibly a small amount of virus production by these tumors. Alternatively, expression of late genes may be controlled, as in productively infected cells, by processes unconnected with methylation. When cells from these tumors were cultured, the extent of methylation of CCGG and GCGC sequences in integrated viral DNA increased.

Only one of three adeno 2-transformed hamster cell lines, all of which contain an intact E_{2a} region, produces the 72K DNA-binding protein encoded in this region (van der Vliet & Levine, 1973; Vardimon et al., 1980; Doerfler, 1984). In the other two cell lines this region is fully methylated, whereas expression is associated with lack of methylation of CCGG sites.

Although we have been rather dogmatic about whether a CCGG site is methylated or not, the published fluorograms seldom give a 100% answer.

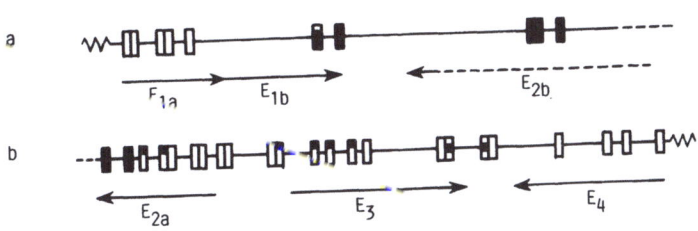

Figure 8.5. Methylation in three adeno 12 transformed hamster cell lines. The boxes indicate the sites investigated (either CCGG or GCGC) at the left (a) or right (b) end of the integrated viral DNA. An open box indicates no methylcytosine A partly or completely filled box indicates methylation in one or all of the cell lines. Arrows indicate the extent and direction of transcription. (From Kruczek & Doerfler, 1982.)

It should be realized that DNA from several million cells is being examined, each one of which may contain up to ten integrated copies of viral DNA. If only half the copies of a particular gene are expressed in half the cells, then a site may appear to be three-quarters methylated and yet expressed. The situation is even more difficult in tumors which, besides the producing cells, contain a variety of other cells, some of which may contain inactive copies of viral DNA.

It is clear that the integrated adeno virus genome provides a perfect example in which gene expression is correlated with the absence of methylation of the promoter-leader and 5' region of the gene. However, as we have said, during productive infection the adenovirus is able to control the sequential expression of early and late genes without the intervention of methylation. It is not, therefore, necessary for the late genes to be methylated for their transcription to be blocked, though the converse may be true; *i.e.*, the absence of methylation in 5' control regions *may* be necessary for transcription. The presence of methylcytosine in the body of the gene appears to have no effect on transcription.

8.3.3b   Retrovirus Proviral DNA

Mouse mammary tumor virus (MMTV) is a type B retrovirus that can gain entry into mice in one of two ways. In many mice (and in all inbred strains), the provirus (*i.e.*, the DNA copy of the viral RNA) is integrated into the germ line DNA in several places, and hence is inherited in the same way as any other gene. The virus may also be transferred in the milk and infect either normal cells or cells already containing endogenous provirus. The genetically transferred proviral DNA is normally inactive and most copies are fully methylated at CCGG sites (Cohen, 1980; Gunzberg & Groner, 1984), though occasionally it can give rise to tumors. When this happens, tumor formation is associated with demethylation of specific proviral sequences in one of the long terminal repeats (LTRs) known to contain the promoter of the 38K protein involved in cell transformation and tumorigenesis (Breznik & Cohen, 1982). The milk-borne virus proviral DNA is undermethylated whether it is present in transformed cells or in cells containing no other proviral DNA. The milk-borne virus proviral DNA is transcribed both in normal and tumor cells, and this leads to production of new virus particles.

In some cell types, including lymphoid cell lines, integrated forms of MMTV proviral DNA can be activated by treatment with glucocorticoids. Cell lines resistant or supersensitive to glucocorticoids have been isolated. Treatment of such cells with dexamethasone for 1.5 cell doubling times does not affect the level of methylation, though transcription of retroviral RNA sequences is induced in the wild type and supersensitive cells. In contrast, the resistant cells that were selected by continuous growth in the presence of high levels of glucocorticoid are already devoid of methyl groups in the 5' LTR and produce retroviral RNA (Mermod *et al.*, 1983). It appears in this case that the observed demethylation only occurs after continuous transcription of the proviral DNA for many generations.

It appears that the state of methylation of MMTV proviral DNA may reflect the chromosomal location of the proviral DNA, and the tissue studied is also very important (Hu *et al.*, 1984; Gunzberg & Groner, 1984). Except for prelactating and lactating mammary gland, normal tissues contain no detectable retroviral RNA, yet proviral DNA from spleen and testis in particular is undermethylated. The undermethylation is not, however, complete, and it is possible that the important sites, *i.e.*, those in the 5' LTR remain methylated. This remains to be investigated.

Mammary tumor virus proviruses integrated into rat hepatoma cells may or may not be activated by glucocorticoid hormones, depending on their site of integration (Feinstein *et al.*, 1982). Nonresponsive provirus has integrated into a region of chromatin insensitive to nucleases, and it is suggested that the surrounding chromatin structure spreads onto the newly integrated provirus to inactivate it. Inactivation could be followed by methylation, but the alternative is also possible—methylation spreads onto the proviral DNA and dictates the chromatin structure assumed. The same proviral DNA integrated into a region of chromatin sensitive to nucleases assumes a sensitive configuration and shows hormone-dependent activation. Similar results have been reported by Gunzberg and Groner (1984), but the results of Jahner and Jaenisch (1985) have been given a different interpretation. These authors conclude that integration of Moloney murine leukemia virus proviral DNA induces a methylation of flanking host sequences which can lead to transcriptional inactivation of host genes.

Chickens contain one or more copies of the proviral DNA of a similar endogenous retrovirus. One such provirus (ev1) is genetically silent and is present in nuclease-resistant chromatin. Groudine *et al.* (1981) have studied a second provirus (ev3) present in a nuclease-sensitive region of chromatin. In particular there are DNase I-hypersensitive sites in the 3' LTR and the 5' LTR indicating that this proviral DNA is transciptionally active. The provirus ev1 is highly methylated at CCGG sequences, but ev3 is totally unmethylated at two such sites and undermethylated at several other sites. When a line of chicken cells (MSB) containing only ev1 is treated with azacytidine for two to three generations there is extensive demethylation of the proviral DNA, a state that is maintained for at least ten generations after removal of the drug. Concomitant with this demethylation, transcripts of ev1 proviral DNA are made and virus particles are detectable.

Similar results have been obtained with murine leukemia proviruses (Hoffmann *et al.*, 1982), and a correlation exists between hypomethylation and transcriptional activity of Rous sarcoma virus proviral DNA in a series of normal and transformed rat cells (Searle *et al.*, 1984). However, in this case, although a DNase I-hypersensitive site is present at the 5′ end of the proviral DNA, in transformed cells the changes in DNA methylation occur in the region of the *src* gene toward the 3′ end of the provirus, far from the site of initiation of transcription. Treatment of phenotypically normal cells with azacytidine leads to the appearance of transformed foci in a dose-dependent manner. Here the correlation between expression and undermethylation is again strong, but not absolute, as proviral hypomethylation is seen in only the minority of cells that become transformed.

8.3.3c   Globin Genes

In mammals and birds hemoglobin contains two types of globin chains—the α type and the β type. Although the genes for these two globin types probably evolved by duplication of an ancestral *globin* gene, they have diverged considerably and each type now consists of several members. In all species one or more globin genes of each of the α and β types are active during development and different genes are active in the adult. Not only are different genes active at different stages of development, but globin is synthesized in different tissues at the different stages.

In the human fetus, zeta (α type) and epsilon (β type) globin genes are active in the embryonic yolk sac. Later in embryogenesis the two α globin genes and the two gamma (γ^A and γ^G) globin genes are active in the liver. In the adult, the α globin genes, the β, and to a lesser extent the delta (β type) globin genes are active in bone marrow cells, as illustrated in Figure 8.6 (Weatherall & Clegg, 1979). The β type globin genes are found adjacent to each other on chromosome 11 and are all transcribed in the same direction. The order in which they are found on the chromosome is the same as the order in which they are expressed during development and also the same as the order in which they developed from their common ancestor (Efstratiadis *et al.*, 1980).

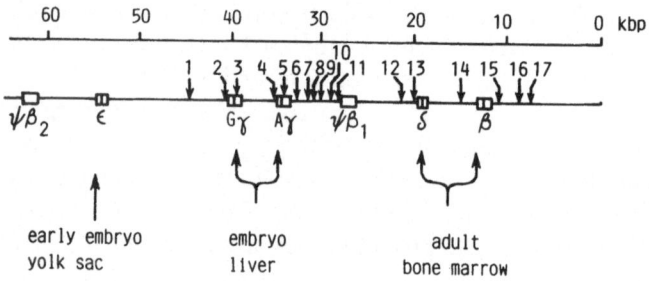

Figure 8.6. The human β-like globin cluster. The boxes indicate the positions of the genes and the arrows indicate the 17 positions probed in the study of van der Ploeg and Flavell (1980).

An examination of 17 sites in the human $\gamma\delta\beta$ globin region failed to show a strong correlation between expression and undermethylation (van der Ploeg & Flavell, 1980) (Figure 8.6). All sites were methylated in DNA from sperm and brain, but site 17 is only slightly methylated in other tissues, irrespective of expression. In fetal liver, sites 2 to 7 are undermethylated (these are in the area of the fetal globin genes), but all sites investigated are undermethylated in placenta or cultured HeLa or KB cells.

The different pathways followed by the hematopoietic stem cells may be influenced in no small way by the tissue environment of the cells, *i.e.,* when in the liver they will produce cells that make fetal globin, but when in the bone marrow they will produce cells that make adult globin. Alternatively, switching may be controlled by a developmental clock (Wood *et al.,* 1985). It is thus little short of amazing that treatment of baboons and man (patients suffering from severe β thalassemia and sickle cell anemia) with 5-azacytidine is reported to activate the fetal globin genes (DeSimone *et al.,* 1982, Ley *et al.,* 1982, Charache *et al.,* 1983). In the latter case, it was hoped that a stimulation of γ globin synthesis would relieve the symptoms caused by the failure to produce β globin. Indeed, after treatment with 5-azacytidine for one week, a substantial increase in γ globin synthesis was detected, associated with hypomethylation of bone marrow DNA in the region of the embryonic and fetal globin genes. It is possible that γ globin genes and a handful of other genes are particularly sensitive to this sort of manipulation. Azacytidine treatment of an erythroleukemic cell line carrying the human β-like globin gene region causes a transient genomic demethylation but within two days most of the DNA is remethylated. The exceptions include the globin genes including the previously inactive γ-globin genes suggesting a selective inhibition of methylation in these areas (Ley *et al.,* 1984). An alternative explanation is that some other toxic function of azacytidine is switching on fetal globin transcription as a byproduct of its primary action (Orkin, 1983). The stem cells may be protected in some way from the toxic effects of azacytidine and may differentiate, as in the fetus, to produce some γ globin-synthesizing erythroblasts, which are now found in the bone marrow.

Shen and Maniatis (1980) carried out a similar analysis of 13 sites around the rabbit β globin cluster. Again, brain DNA was fully methylated at most sites. In the fetus, the yolk sac blood island DNA is only partially methylated at two sites that are fully methylated in DNA from all nonexpressing tissues. Little difference is found with five other sites. All sites around the adult β globin genes are partially methylated in all tissues, with the exception of one site that is fully methylated in DNA from kidney and brain. These results are not very clearcut, and this may be for one of three reasons. First, there may be no relationship between methylation and gene expression. Second, the rate of globin synthesis may vary dramatically from one erythroid cell to another, and some transcription of globin genes may even occur in nonerythroid cells. Third, the tissues used contain several cell types, and even the brain used may be contaminated with erythroid cells. Any of these problems would cloud the issue.

In murine erythroleukemia cells, globin synthesis can be induced by a variety of agents such as dimethyl sulfoxide or ethionine or hexamethylene bisacetamide. Even in uninduced cells the globin chromatin shows enhanced sensitivity to nucleases; however, certain hypersensitive sites appear near the 5' end of both the α and β globin genes following induction (Sheffery et al., 1982). No change in the methylation pattern is seen and most sites investigated are already undermethylated in uninduced cells. The changes in DNase I sensitivity are apparent at 48 hours after addition of the inducer, and cytoplasmic globin mRNA is certainly detectable before this. Within this time scale there is no opportunity for a cell to divide twice to generate an unmethylated from a fully methylated site, and hemimethylated sites are not detectable by HpaII treatment (but see Section 8.4.3a). It remains possible but unlikely that hemimethylated sites have more significance than we have credited them with.

The chicken globin genes show a much clearer relationship between undermethylation and expression (Weintraub et al., 1981; Groudine & Weintraub, 1981; McGhee & Ginder, 1979; Haigh et al., 1982). There are two α globin genes (α^D and α^A) which are transcriptionally active in adult and in 14-day definitive red blood cells. A third α globin gene is not active after about five days of development. In active chromatin from 14-day red blood cells, the adult α globin genes are sensitive to DNase I from a point at the 5' side of the first α gene, through a 1500-bp spacer and the second α gene to about 1500 bp on the 3' side of the second gene. Most, if not all, of this region is transcribed, and all sites tested are unmethylated (Weintraub et al., 1981). The embryonic α globin gene region of chromatin shows only limited DNase I sensitivity at this time, and the CCGG sites tested are methylated. However, at five days two CCGG sites bordering the embryonic gene are unmethylated. DNase I-hypersensitive sites are also present in chromatin near the 5' end of active genes.

Haigh et al. (1982) looked in addition at the methylation of HhaI sites (GCGC) in this region and found the correlation between expression and lack of methylation to be not perfect. Three HhaI sites showed partial methylation, which varied between tissues in a manner apparently unrelated to gene activity, as shown in Figure 8.7. All sites were completely methylated in sperm and almost completely methylated in brain DNA. Sites M2, M3, and M5 (and to a lesser extent M4) are resistant to cleavage with MspI in all tissues tested. This may imply a methylation of the 5' cytosine in the sequence CCGG, but has been shown in some cases to be caused by the inability of MspI to cut the sequence CmCGGCC (Busslinger et al., 1983a).

In the region 5' to the chicken adult β globin gene, there are 33 potentially methylatable CG dinucleotides in a region that is nucleosome-free in

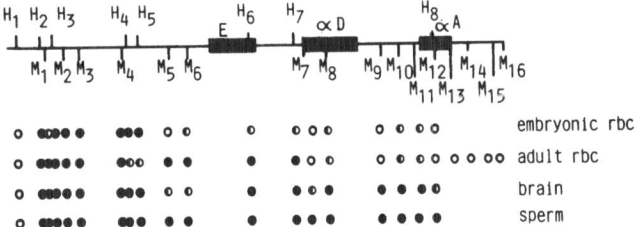

Figure 8.7. Methylation of the chick α globin gene region (based on Haigh *et al.*, 1982, and Weintraub *et al.*, 1981). Filled circles represent complete or almost complete methylation and open circles represent little methylation. Half-closed circles represent intermediate values. H1 to 8 are HhaI sites and M1 to 16 are MspI (HpaII) sites. The filled boxes represent the three α globin type genes: Embryonic (E) and Adult (α^D and α^A).

definitive red cells but not in primitive erythrocytes of five-day embryos or other nonexpressing cells (McGhee *et al.*, 1981). This region has been cloned into a plasmid, but methylation of 22 of these sites with HpaII methylase has no discernible effect on DNA structure. Thus *in vivo* methylation of this region is unlikely to have any long-range effects, such as induction of Z-DNA conformation (Nickol & Felsenfeld, 1983). Although some of thee sites are unmethylated in cells that are expressing, or have expressed, the adult β globin gene (adult reticulocytes and erythrocytes), they are also partly unmethylated in a variety of tissues that are not and have not expressed this gene (*e.g.*, oviduct, brain, and embryonic red blood cells) (McGhee & Ginder, 1979).

It is easy to overinterpret these results and claim a good correlation between undermethylation and gene expression, but there are too many partial or complete exceptions to the rule for more than limited confidence to the placed in this simple idea.

It is possible that a particular cell synthesizing fetal globin changes over and starts to make adult globin. It is more likely, however, that the stem cell differentiates in such a way that instead of dividing to form cells capable of making embryonic or fetal globin, it now follows a different pathway to produce cells that will synthesize adult globin.

It seems possible that different methylation patterns may exist in different tissues for a variety of reasons, but that expression of a gene is always correlated with undermethylation of only a few particular sites.

8.3.3d   Ovalbumin Gene

The immature chick oviduct does not make ovalbumin unless hormonally stimulated, in which case ovalbumin becomes the major protein synthesized by the duct cells. In the chromatin of expressing tissues, there are DNase I-hypersensitive sites upstream from the gene that are present to a lesser extent in nonexpressing tissues (Kaye *et al.*, 1984). A number of CCGG sites have been investigated for their state of methylation. Some of these sites are always methylated and some are never methylated, irrespective of the activity of the gene. However, some sites show a variable level of methylation, being undermethylated in expressing tissues (Mandel & Chambon, 1979; Kuo *et al.*, 1979). The changes at the variable sites are far from complete, but this is caused largely by the fact that the oviduct is a mixed-cell tissue, and only some of the cells are making ovalbumin. Taking advantage of the fact that active chromatin is preferentially sensitive to DNase I, Kuo *et al.* (1979) showed that the variable sites are, in fact, totally unmethylated in the active, DNase I-sensitive chromatin from duct cells. The converse is, however, not true, as even in erythrocyte nuclei these sites are not fully methylated. Does this imply a certain level of transcription of the ovalbumin genes in erythrocytes? Or is the pattern of methylation a consequence of the chromatin structure, where proteins bound to the DNA may interfere with methylation while not necessarily inducing transcription?

8.3.3e   Metallothionein Gene

Metallothioneins are small, cysteine-rich proteins that bind heavy metals such as cadmium and zinc and are involved in detoxification. Transcription of the gene for metallothionein I is induced by heavy metals or by glucocorticoid treatment in normal mouse cells, but Mayo and Palmiter (1981) have isolated T cell lines (the W7 thymoma and the S49 lymphoma) that are rapidly killed by cadmium. Although the gene for metallothionein is present in these T cell lines, it is not transcribed when the cells are challenged with heavy metals or glucocorticoids. In cells that can express metallothionein, all the HpaII (CCGG) sites in the vicinity of the gene are unmethylated, but they are all methylated in W7 cells. It is not clear at what point the block in induction exists. It could be that these cells fail to make a carrier protein (receptor) that is necessary to take the inducer to the gene.

    Treatment of W7 cells with 5-azacytidine causes the metallothionein gene to become inducible by cadmium and by glucocorticoids, and resistant W7 clones now are unmethylated at HpaII sites within and flanking the gene (Compere & Palmiter, 1981). In fact, the metallothionein gene has become amplified by the azacytidine treatment followed by selection in cadmium, and it appears that it is only the amplified copies that are unmethylated as a certain proportion of the DNA still remains resistant to HpaII.

    What is particularly interesting in this system is the time course of reactivation of the gene. The maximum amount of metallothionein mRNA per cell is found 17 hours after commencement of azacytidine treatment, though half maximum levels are reached in about 10 hours. (It is confusing in this experiment that the inducer, cadmium, was added 8 hours prior to harvesting.) As the doubling time of these cells is about 9 hours, it appears that it is only when cells pass through S phase for the *second* time that they become inducible, *i.e.,* two rounds of replication are required to render the gene inducible. This would allow time for unmethylated DNA to be produced by replication in the absence of methylation.

This was not the conclusion of Compere and Palmiter (1981), who interpret their results as showing that the DNA becomes expressible when it is hemimethylated, but this is not clear from their data.

Continued growth of azacytidine-treated cells in the absence of the drug causes a gradual loss of inducibility; whether this is because uninducible cells (which have not incorporated azacytosine into their DNA) grow faster and replace the inducible population or because the demethylated gene becomes remethylated and inactivated cannot be determined.

Cadmium-resistant S49 lymphoma cells are obtained several days after treatment with ultraviolet irradiation. During this time no selection is applied, but several rounds of DNA synthesis occur. (Resistant cells cannot be detected by growth on cadmium-containing medium.) The resistant cells investigated by Lieberman *et al.* (1983) showed extensive demethylation of one of the alleles for metallothionein I, and some cells also showed demethylation of a β globin gene. It is possible that the demethylation was brought about by extensive repair of DNA following irradiation at a time when DNA methylase was deficient, though the area of the gene demethylated was substantial (at least 2500 bp).

8.3.3f   Immunoglobulin Genes

Immunoglobulins are made up of heavy and light chains, each of which consists of a part in which the amino acid sequence is constant in all lymphocytes of a given type joined to a region in which the amino acid sequence varies from cell to cell. Comparison of the constant regions has allowed several types or classes of immunoglobulins to be defined.

The light chains are synthesized from genes which, in the germ line, are made up of three widely separated sequences (V, J, and C). The V and J sequences are linked together in functional lymphocytes. The heavy chains are similarly constructed from four sections (V, D, J, and C), but during maturation of B lymphocytes, a switch occurs and the constant heavy region expressed changes, and this leads to production of different classes of antibody. The first immunoglobulin to be expressed (IgM) is synthesized on a gene where DNA rearrangement has produced a section VDJ$_H$-Cμ. Later, the Cδ gene is expressed by transcription through a rearranged μ gene in a VDJ_H-$C\mu$-$C\delta$ transcription unit, and finally the $C\mu$ and $C\delta$ regions are deleted, bringing the $C\gamma$ region close to the VDJ_H region. The current concept about the control of transcription of immunoglobulin genes has been summarized by Perry (1984).

Rogers and Wall (1981) examined the methylation of the constant region of the heavy chain genes during B cell differentiation. They used double digests of nuclear DNA with EcoRI and either MspI or HpaII and probed the Southern blots with a labeled cloned fragment from near a reference EcoRI cut, as shown in Figure 8.8. In this way they obtained a single band from the MspI double digest and a ladder from the HpaII double digest, which indicates the degree of methylation of a number of adjacent sites (Figure 8.8). Of course, if the rightmost CCGG site is totally unmethylated, only the 4.1-kbp band will be seen in an HpaII digest, and no evidence can be obtained about the degree of methylation of the other sites.

In liver the $C\mu$, $C\delta$, and $C\gamma_1$ genes are all largely (but not completely) methylated, whereas in expressing cells they are largely unmethylated. This conclusion comes from a study of the results diagrammed in Figure

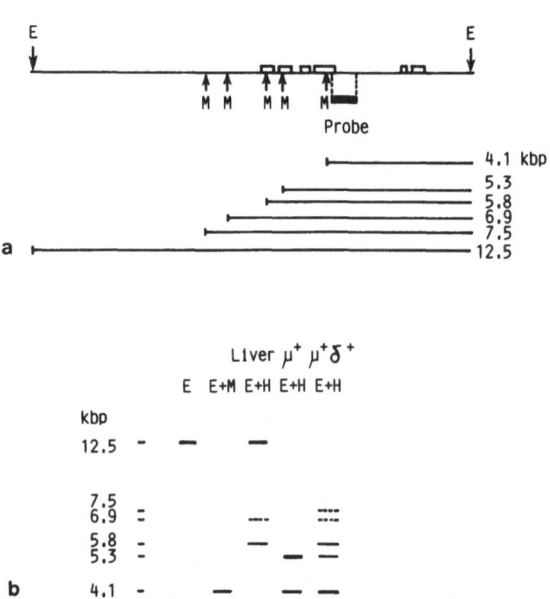

Figure 8.8. Methylation of immunoglobulin μ gene. (a) The region of the gene is shown between two EcoRI sites. Exons are shown by boxes, and the arrows below the line indicate MspI sites. The location of the probe used is indicated, and the sizes of possible fragments are delineated. (b) Southern blots (diagrammatic) indicate that in liver most of the CCGG sites are methylated, as the major band is 12.5 kbp. In two cell lines expressing μ, the 4.1 kbp band is strong, showing the rightmost CCGG to be largely unmethylated. (Simplified from Rogers & Wall, 1981.)

8.8. Most of the liver DNA is present as the 12.5-kbp band produced by the two EcoRI cuts, though some molecules are unmethylated at the 5.8-kbp site, and a very small number at the 6.9-kbp site. In cells expressing IgM, all μ genes are unmethylated at the 4.1-kbp site and/or the 5.1-kbp site. We have no information about the other sites. In cells making IgM and IgD, about a third of the μ genes are unmethylated at the 4.1-kbp site, and the rest have at least one unmethylated site.

In one cell line where two alleles of the μ gene are present, both are demethylated although it is known that only one is transcribed. In the Cδ gene one site is always methylated in expressing and nonexpressing tissues.

Similar results have been obtained by Storb and Arp (1983) but, Akira *et al.* (1984) found that the Cμ genes are undermethylated in a series of transformed cell lines, whether or not they are expressed, and they also identified one B cell clone that was expressing the Cμ gene even though it was heavily methylated.

In pre-B cell lines the $V_H D_H J_H$ is undermethylated but only low levels of transcription occur. The Cμ region is fully methylated. Stimulation with lipopolysaccharides or by fusion with a plasmacytoma cell leads to elevated transcription of the Cμ gene but only in the latter case does demethylation of the Cμ gene occur (Gerondakis *et al.* (1984). This was interpreted as showing that demethylation stabilizes a transcriptional program set by other factors.

Storb and Arp (1983) showed that in liver DNA the light chain genes are largely or completely methylated at the few sites (HpaII and HhaI) investigated, and most are undermethylated in expressing tissues. These conclusions seem rather tentative as very few sites have been investigated and some of these are more than 4 kbp away from the coding region. A more discriminating study by Mather and Perry (1983) showed that the constant region (C$_\kappa$) is always hypomethylated in germ line and in expressing tissues, and this correlates with its DNase I hypersensitivity. However, the variable regions are highly methylated in the germ line and are not so susceptible to DNase I. On fusion to the κ locus, a variable region becomes DNase I sensitive and hypomethylated. This change, although extending several kilobases upstream from the transcription initiation site, is restricted to the single Vκ allele undergoing fusion.

8.3.3g   Summary

On reading the abstracts of the papers commented on here as well as of others coming to the same conclusion, there appears to be strong support for the idea of the correlation between undermethylation and expression. However, on closer examination of the data presented, it is clear that relatively few sites have been investigated and there are so many exceptions, both in expressing and nonexpressing tissues, that one is left in doubt as to the significance of the findings. What is clear is that all of the genes studied (see also the next section) are completely or almost completely methylated in sperm and only somewhat less so in brain. Perhaps in these tissues, large stretches of the genome are inactive and heterochromatic, and this is associated with higher levels of methylation (except for satellite DNA in sperm, see Section 7.3.2). An exception to the finding of high levels of methylation in sperm DNA comes from the study of the genes for adenine phosphoribosyl transferase and dihydrofolate reductase (Stein *et al.*, 1983). These genes code for housekeeping proteins and are presumably active in sperm. Using CpG restriction enzymes, they were found to be heavily methylated throughout the gene with the exception of their 5' regions in all tissues tested. The α2-I collagen gene is also undermethylated at its 5' end in sperm DNA, but in this case there is no evidence for activity in this tissue (McKeon *et al.*, 1982, see Section 8.3.4b). The thymidine kinase gene, although inactive in mature sperm, is active in early spermatogenesis and is reactivated following fertilization and always retains one unmethylated site (Groudine & Conkin, 1985; see Section 12.6). Such inferences have led to the conclusion that undermethylation is *necessary but not sufficient* for gene expression.

8.3.4 Genes that Do NOT Show a Correlation Between Expression and Undermethylation

It is clear from the preceding section that any relationship between undermethylation and gene expression is far from perfect, and this is borne out by studies on several other genes listed in Table 8.2. Some of these we shall now look at in more detail.

8.3.4a   Vitellogenin Genes

In *Xenopus* vitellogenin synthesis occurs in female liver and can be stimulated in males and in isolated male hepatocytes by treatment with estrogen. In culture, vitellogenin mRNA can be detected within 3 hours of exposure of hepatocytes to the hormone, and the maximum rate of synthesis occurs within 6 hours (Searle & Tata, 1981). This is in normal cells which have never previously made vitellogenin and which, *in vivo*, would never normally make vitellogenin. Gerber-Huber *et al.* (1983) investigated the methylation of two of the *Xenopus* vitellogenin genes (A_1 and A_2). There are four genes in *Xenopus*, and all four are expressed simultaneously to equal extents. On estrogen stimulation of hepatocytes the chromatin of the A_1 and A_2 genes becomes sensitive to DNase I treatment, but no changes are seen in the methylation of the DNA (Gerber-Huber *et al.*, 1983; Folger *et al.*, 1983). In the A_2 gene 20 HpaII or HhaI sites are fully methylated in untreated and hormonally stimulated hepatocytes and in erythrocytes. In the A_1 gene the same is true for 12 sites, while a single site is hypomethylated in all three situations. The sites investigated stretched from the 5'-flanking region, through the central part of the gene to the 3' end (the single hypomethylated site was toward the 3' end of the A_1 gene).

A similar situation exists with the chick vitellogenin gene (Wilks *et al.*, 1982; Meijlink *et al.*, 1983). Normally vitellogenin is only made in the liver of female chickens, but administration of estradiol to a rooster leads to vitellogenin synthesis in the liver, which starts within 6 to 10 hours of treatment and reaches a maximum after three to four days and then falls back to control levels. Cell proliferation is not required. The vitellogenin gene is heavily methylated in all tissues studied except the liver of the laying hen, where several HpaII sites are unmethylated, as indicated in Figure 8.9. A small amount of demethylation occurs toward the 5' end of the gene in hormonally treated roosters. This begins at about 10 hours and is still only partially complete at three days. The demethylation persists and is still present at 32 days, when vitellogenin synthesis has long ceased. This site (HpaII-1) is also demethylated in chick oviduct cells that do not synthesis vitellogenin, but it is not in fact essential at all for vitellogenin synthesis. Some birds have been found that homozygously lack the site completely (Burch & Weintraub, 1983). There are changes in chromatin structure that accompany the induction of vitellogenin synthesis in hormonally treated male chicken liver cells. Even prior to treatment,

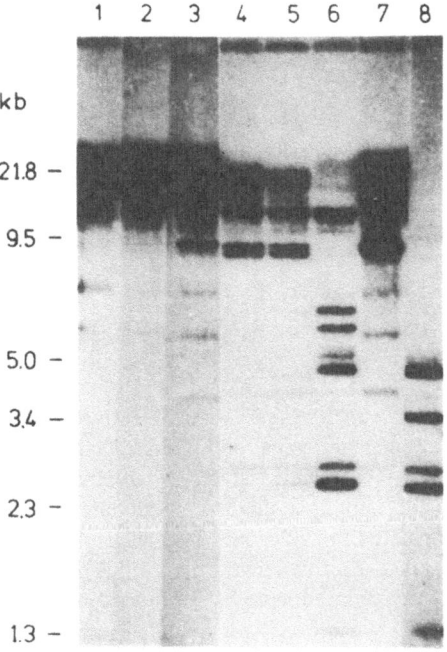

Figure 8.9. Methylation of the chick vitellogenin gene. Liver DNA from chicks was digested with HpaII and the Southern blot probed with ^{32}P-labeled vitellogenin DNA (1) embryo; (2) rooster; (3) and (4) roosters 3 and 32 days after a single estradiol injection, respectively; (5) a rooster after repeated estradiol injections; (6) a laying hen; (7) oviduct DNA from a laying hen; (8) MspI digest of liver DNA from an estrogen-treated rooster. (From Meijlink *et al.*, 1983.)

there are DNase I-hypersensitive sites present and three additional ones appear near the 5′ end of the gene. Two of these are stable and persist after vitellogenin synthesis has ceased, but the third (located 700 bp upstream from the start of transcription) is only present in liver chromatin during estrogen treatment. This hormone-responsive site appears before the HpaII-1 site is demethylated and is also present in the oviduct (another estrogen-responsive tissue). It appears that estrogen induces changes in the chromatin of target tissues that may or may not lead to transcription of the vitellogenin genes but do result in the production of DNA unmethylated at specific sites (Burch & Weintraub, 1983).

Although the demethylation observed starts only after transcription has been induced and continues long after transcription has terminated, rigid conclusions are not possible because even in the untreated rooster, all HpaII and HhaI sites are methylated to only 90 to 95%. The following argument shows that it is possible for the 5 to 10% of cells with an unmethylated HpaII-1 site to be those that respond initially to the hormone treatment that also induces cell division in about 40% of the cells. Such replication may amplify the unmethylated fraction to 55% of the population within 48 hours, which is consistent with the level of methylation seen, as shown in Figure 8.10.

This possibility appears to be ruled out, however, by the more recent experiments of Wilks *et al.* (1984). They first confirmed that transcription of the chick liver vitellogenin gene is induced by estrogen before demethylation of the HpaII site 611 bp 5′ to the start of transcription, as Figure 8.11 illustrates. They further showed that both of these processes (tran-

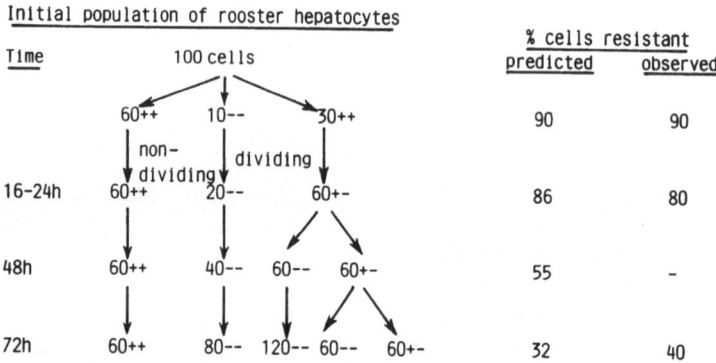

Figure 8.10. The effect of selective cell division on the size of the HpaII-resistant vitellogenin gene population. Genes fully methylated (++), hemimethylated (+−), or unmethylated (−−) at HpaII-1 are shown. It is suggested that 10% of the cells initially contain unmethylated genes, and that these are among the 40% selected for division. Methylation may be blocked at HpaII-1 following division, and so after two and three rounds of replication the population of cells containing unmethylated genes would rise to 45 and 68%, respectively.

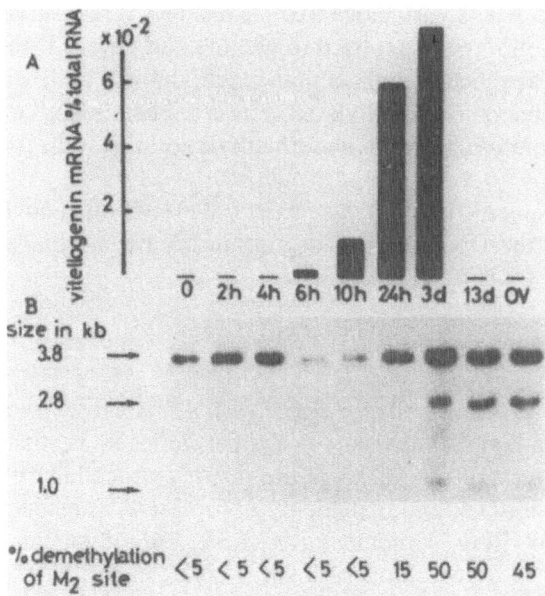

Figure 8.11. The time course of vitellogenin mRNA induction and demethylation at site M2 of the chick vitellogenin gene. Panel A shows the time course of mRNA accumulation at various times after estrogen administration. In panel B, DNA from each time point was digested with HpaII and BamHI, and the Southern blot probed with a ^{32}P-labeled cloned gene. Bands at 2.8 and 1.0 kbp indicate hypomethylation at the M2 site. The percent demethylation is calculated from a densitometry trace: (intensity of bands 2.8 + 1.0) × 100/(intensity of bands 3.8 + 2.8 + 1.0). (From Wilks *et al.*, 1984.)

scription and demethylation) occur in the presence of inhibitors of DNA synthesis. The inhibitors used, hydroxyurea and cytosine arabinoside, produced 90 to 94% inhibition of DNA synthesis over a period of five days. This is bound to have had some serious side effects, but these did not include induction of vitellogenin gene transcription in the absence of hormone treatment.

One explanation for the mechanism of generating an unmethylated region of DNA in the absence of DNA replication involves the action of a site-specific demethylase. Although demethylase activity has been reported to be present in Friend erythroleukemia cells (Gjerset & Martin, 1982), we have been unable to confirm their findings (Adams *et al.*, unpublished data). The activity of a site-specific demethylase would, however, be undetectable by any of the methods used, and so this possibility cannot be ruled out.

An alternative explanation, which is described more fully in Chapter 12, depends on the exposed nature of the 5' region of active genes. The demethylated site is very close to a DNase I-hypersensitive site that may *in vivo* be continuously exposed to nicking and repair. If this repair takes place in the absence of concomitant methylation, both strands of DNA may quickly become demethylated. If this explanation is correct it implies that the correlation between undermethylation and transcription is fortuitous.

Even if a specific demethylase exists, its function is unclear because it acts only on the DNA once transcription has been initiated.

8.3.4b   Collagen Genes

Five distinct types of collagen, each with a distinctive tissue distribution are found in higher eukaryotes. Type I collagen is made up of three chains; two $\alpha 1$ and one $\alpha 2$. It is the major extracellular protein of skin, bones, and tendons, and is made by fibroblasts. Transformation of chick fibroblasts by Rous sarcoma virus (RSV), or of human fibroblasts by SV40, leads to a switch-off of Type I collagen synthesis. The gene for $\alpha 2(I)$ collagen is extremely long (38 kbp) and contains more than 50 short exons, each containing a multiple of nine base pairs corresponding to the three-amino-acid repeating structure of the protein.

In erythrocytes and sperm (which make no collagen), brain (in which the glial cells make small amounts of type I collagen), and in normal and RSV transformed chick fibroblasts, the methylation pattern of the $\alpha 2(I)$collagen gene is the same when measured by sensitivity to a number of methyl-responsive restriction enzymes (McKeon *et al.*, 1982). All six sites investigated around the 5' end of the gene are completely unmethylated. Six sites in the middle and 3' end of the gene show marked or complete methylation, which does not vary between the tissues investigated. In contrast, some increased methylation of the type I $\alpha 1$ and $\alpha 2$ collagen genes is seen in SV40 transformed human fibroblasts (Parker *et al.*, 1982).

The chromatin containing all regions of the chick $\alpha 2(I)$ collagen gene is equally sensitive to nuclease treatment. In addition, there is a DNase I-hypersensitive site 100 to 300 bp 5' to the start of transcription, which is only present in active cells, though it is present in the transformed cells where the rate of transcription is markedly reduced (McKeon *et al.*, 1984). Thus, although changes in chromatin structure accompany changes in gene expression, there is no evidence for any corresponding changes in methylation.

8.3.4c   Insulin Genes

There are two insulin genes in the rat, differing only in noncoding sequences. Both genes are active in insulin-producing tissues but to differing extents. In one solid tumor in which the ratio of insulin I RNA to insulin II RNA is 0.8, the insulin I gene is not methylated at six sites and only slightly methylated at a seventh, whereas the insulin II gene is much more highly methylated at all sites (Cate *et al.*, 1983). The situation is complicated by the fact that most of the undermethylated sites in the insulin I gene are absent in the insulin II gene, but this at least shows they are not essential for control of expression. In the region of homology 5' to the gene, the insulin II gene has only one CG dinucleotide, and the insulin I gene has five, none of which correspond to those in the insulin II gene. In nonexpressing tissues, the insulin I gene is fully methylated, whereas the insulin II gene is partly undermethylated at some sites. Even in brain DNA, four sites are methylated to only 89% (brain DNA is usually heavily methylated), and the same sites are only 18 to 21% methylated in liver DNA.

In the human insulin gene there is a CG dinucleotide about 100 bp from the 3' end of the gene that shows an unstable methylation pattern, *i.e.*, the extent of methylation in a given tissue varies from individual to individual. The level of methylation can be low in white blood cells, which do not synthesize insulin (Ullrich *et al.*, 1982).

8.3.4d Albumin and α Fetoprotein Genes

The α fetoprotein is a major constituent of fetal serum, and it is synthesized in the fetal liver and yolk sac. After birth it is replaced by albumin, but it can reappear when certain liver tumors are present. The two proteins are structurally similar and have considerable amino acid homology, and in the mouse the two genes are adjacent to each other on chromosome 5 (Ingram *et al.*, 1981). Switching in expression occurs in liver parenchymal cells during a few divisions at the late fetal stage, although the albumin gene is actively transcribed in the rat fetal liver by 18 days. At this time, when both genes are being transcribed, they are both heavily methylated. In the adult rat liver, the level of methylation of both genes falls, yet only the albumin gene is active, *i.e.*, the α fetoprotein gene shows a *direct* correlation between methylation and expression (Kunnath & Locker, 1983; Vedel *et al.*, 1983). This result is in direct contrast to that found in the mouse, where there is an inverse correlation between expression and methylation of six HpaII sites in the α fetoprotein gene (Andrews *et al.*, 1982).

Expression of the rat albumin gene has also been studied in a series of clones of rat hepatoma cells, which make albumin at widely differing rates. The initial cell line (H411 EC3) stably expresses a group of specific liver functions, including the secretion of albumin and the synthesis of enzymes involved in gluconeogenesis. It has been subcloned a number of times and the morphology and biochemistry appear stable. One such subclone is Fao. However, by selecting against cells that grow in glucose-free medium, and by other techniques, some subclones (4C, 5C, 2A, H5) have been isolated, which have lost these differentiated functions, but revertants (dag) can be isolated from these at very high frequency (2–12%) by growth at high cell density (Deschatrette, 1980). In addition to these cell lines, Ott *et al.* (1982) studied the methylation of the albumin gene in cell hybrids between nonproducing variant cells (H5) and the differentiated parent cells. HF1 and HF8 are such hybrids that retain all 106 chromosomes. In all cases, the amount of albumin present in a cell is directly proportional to the amount of albumin mRNA or nuclear RNA present, showing that control is exerted at the level of transcription. There are five MspI sites in the rat albumin gene; four of these were studied. The first is in the first of 15 exons, and the last is in the last intron, as shown in Figure 8.12. In rat liver all sites are methylated in only 10 to 30% of cells. Since rat liver is about 80% hepatocytes, these results confirm those noted above that the albumin gene is unmethylated along its length in adult rat liver. However, in the hepatoma clones H411 and Fao, sites 3, 4, and 5 are heavily methylated despite the finding that all cells synthesize albumin (Figure 8.12). Site 1 is, however, completely unmethylated. Thus undermethylation of site 1 alone correlates with gene expression. Although clones 4C, 5C, etc. were selected for absence of two liver *enzymes,* they have lost expression of the entire spectrum of liver functions including albumin synthesis. Each of these clones is methylated at site 1. In clone 3A only half the alleles are methylated, yet none synthesize albumin, so the methylation cannot be the factor blocking expression.

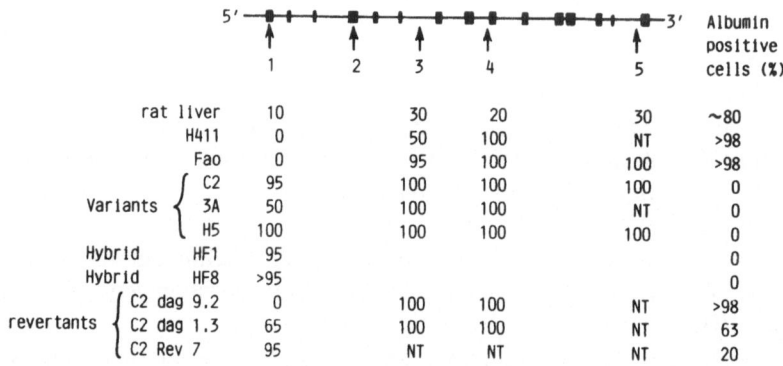

Figure 8.12. Methylation of the rat albumin gene in liver and various clones of cultured cells. The top line indicates the gene with the exons depicted as black boxes. The positions of five MspI sites are indicated by arrows. Under each site is indicated the percent methylation for DNA from the various cells indicated in the left-hand column. The right-hand column indicates the proportion of cells synthesizing albumin. (From Ott *et al.,* 1982.)

In hybrid cells obtained by fusing Fao and H5 cells, albumin is not made, *i.e.*, the dedifferentiated state is dominant, and hence differentiation cannot depend on some factor that may be lost from the nonproducing variants. Site 1 in HF1 and HF8 hybrid cells is almost completely methylated, implying *de novo* methylation at this site following fusion. Strangely, Drahovsky *et al.* (1980) have shown the overall levels of methylation to rise following fusion of two mouse cells (A9 and Ehlich ascites cells). The proportion of cytosines methylated in the hybrids ranged from 4.44 to 6.78. No changes in DNA methylation were found on fusing mouse and human cells, however (Feinstein *et al.*, 1985a).

The dag revertants of clone C2, which arose by growth at high cell density, are generally unmethylated at site 1 and 100% positive for albumin synthesis (Figure 8.12), but one (C2 dag 1–3) is heterogeneous. On subcloning, some heterogeneous subclones were again obtained, indicating considerable instability with regard to expression of albumin and methylation at site 1. Revertants obtained at low frequency by growth in glucose-free medium (*e.g.*, C2 Rev7) are also heterogeneous, and in C2 Rev7 cells, site 1 is 95% methylated yet 20% of the cells synthesize albumin, *i.e.*, methylation of site 1 does not preclude gene expression.

8.3.4e   Summary

It is obvious from these results that the albumin gene is extensively methylated and still expressed in all cells other than liver tissue. Only a small region of the genome (that around site 1) shows any correlation between undermethylation and gene expression, but as the amount of albumin synthesized per liver cell is much greater than in the cultured cells, there may be a correlation between *frequency* of transcription and undermethylation. As genes undergoing very frequent transcription are known to show a breakdown of chromatin structure (Karpov *et al.*, 1984; Spadafora & Crippa, 1984), it is possible that this interferes with methylation, resulting in the lower levels of methylation associated with transcribing regions.

Where the inverse correlation between transcription and methylation completely breaks down, as with the histocompatibility antigen H-2κ gene in teratocarcinoma cells (Tanaka *et al.*, 1983), such a simple explanation cannot hold.

Two housekeeping genes (adenine phosphoribosyltransferase-*aprt* and dihydrofolate reductase-*dhfr*), which are active in all cells, are heavily methylated throughout, except at their 5' ends where they are completely unmethylated (Stein *et al.*, 1983). This is true even in sperm, and hence one constant methylation pattern is maintained throughout development. The presence of control proteins bound to sites at the 5' ends of genes may interfere with methylation over quite small regions. These proteins may themselves block accessibility to DNA methylase or, as suggested for the vitellogenin gene, they may generate nucleosome-free hypersensitive regions in which the DNA is constantly turning over. Such turnover would rapidly generate unmethylated DNA if it occurred in the absence of concomitant methylation, as it might in non-S-phase cells where DNA methylase is limiting (see Chapter 12). The methods used to detect unmethylated regions rely on the location of particular restriction enzyme recognition sites in the sensitive regions. Should these not occur, then the expected correlations will not be found, and irrelevant changes at other locations may lead to misinterpretation of the results.

These problems are discussed further in Chapter 12.

8.4 Methylation and Expression of Genes Transferred into Animal Cells

Recently several procedures have become available that allow the experimental introduction of exogenous DNA into eukaryotic cells. This DNA transfection procedure has enabled the negative correlation between methylation and the expression of the introduced gene to be studied in more detail. In particular, studies of the effect of selective *in vitro* methylation of the introduced gene sequence on its subsequent expression may help to solve the dilemma as to whether transcription blocks methylation or the reverse. In addition, transcription can be studied in cell-free systems, but such *in vitro* transcription systems (*e.g.*, Jove *et al.*, 1984) have so far shown no response to varied levels of DNA methylation. They also have proved to be of only limited success in the study of control sequences in general.

Perhaps the easiest approach is to allow the recipient cells to take up added DNA adsorbed onto particles of calcium phosphate (Wigler *et al.*, 1979). Once inside the cell, the various DNA molecules appear to be joined together into little packages, and later some of these are integrated at random sites into the recipient cells' chromosomes. Very soon after uptake, the foreign DNA can be expressed as RNA and protein leading to *immediate expression* assays. The reproducibility of these tends to be rather variable, and, as the aim is to obtain high levels of expression, large amounts of DNA are required to ensure that most cells become transfected. It is common in such assays to compare the expression of two genes (one a control and the other the test gene) that are cotransfected.

Alternatively, expression can be studied in clones selected from the transfected population. In this case the most commonly transfected gene is the thymidine kinase gene of herpes simplex virus (HSV *tk*), which produces TK^+ transformants of TK^- mutant mouse or hamster cells. The transformants can be selected by growth of the cells in HAT medium (medium containing hypoxanthine, aminopterin, and thymidine), which permits the growth of only those cells containing functional thymidine kinase and hypoxanthine phosphoribosyl transferase genes. Of course, in this type of experiment the transfected gene has to be present in the transformed cell for many generations before the analyses can be carried out, giving ample time for changes in methylation patterns to occur (see below). Furthermore, if the gene of interest is that which is being selected for, it is, of necessity, expressed, and so a better experiment is to cotransfect with a second gene as well as the HSV *tk* gene, select with HAT for clones containing the latter, and then look at the expression and methylation of the second gene.

In addition to introducing DNA into cells on a calcium phosphate precipitate, the DNA can actually be injected directly into cells. This is much easier with the giant oocytes of *Xenopus,* but is also giving interesting results with tissue culture cells and cells of early mouse embryos. It has become clear that by whichever means the foreign DNA is introduced, the nature of the recipient cell is important for both methylation and expression. As we report in Section 8.5.6, DNA injected into very early mouse embryo cells or teratocarcinoma cells is rapidly methylated *de novo,* but this seems to be only a very infrequent reaction in other tissue culture cells or *Xenopus* oocytes. However, expression of introduced DNA depends on factors present in the recipient cell and immunoglobulin genes are only expressed in the homologous myeloma cells and β globin genes are poorly expressed in all but erythroleukemia cells (Stafford & Queen, 1983; Chao *et al.,* 1983). The presence of cell-specific factors that may act independently of (and possibly in addition to) methylation must be borne in mind when trying to interpret the results of transfection experiments.

8.4.1   Stability of Methylation Patterns

The DNA transfected into eukaryotic cells has normally been obtained by cloning techniques. As these mostly involve growth of the gene sequence as part of a plasmid or phage in an *E. coli* host, the input DNA lacks the methylcytosines normally associated with eukaryotic DNA. This is also true of the prokaryotic DNA surrounding the cloned gene and also applies to DNA from eukaryotic viruses, *e.g.,* the HSV *tk* gene. When such unmethylated DNA is introduced into cultured mouse cells or rat liver cells it remains unmethylated and clones can be isolated after many generations in which the DNA is still largely susceptible to cleavage with HpaII (Wigler *et al.,* 1981; Pollack *et al.,* 1980; Stein, Gruenbaum, Pollack, Razin & Cedar, 1982).

Looking at this more critically, however, the results show that after about 50 generations, a particular HpaII site does occasionally become methylated. Let us say there is a one in 100 chance of any unmethylated CpG becoming methylated *de novo* in 50 generations. By this time the original cell has divided to form about 10^{15} cells, and to be visible as a methylated HpaII site the methylation must be present in about 10^6 cells, *i.e.*, the *de novo* methylation event probably occurred in one in 10^9 cells. As there are about 4×10^7 *CpGs per mammalian cell, this implies that only one de novo* methylation occurs in every 25 cells per day. Such *de novo* methylation does not appear sufficient to methylate integrated viral DNA in transformed cells (Section 8.5.5), but in this case, it is possible that there is a positive selection for cells in which the late viral genes are inactivated by methylation.

Rather than transfecting cells with unmethylated DNA, it can first be treated *in vitro* with the HpaII methylase. This bacterial enzyme will methylate the internal cytosine in the sequence CCGG (see chapter 5). When such DNA is introduced into mouse cells, most of the methyl groups are lost quite rapidly but in a fairly random manner. In those clones that become established in which the transfected DNA retains methyl groups these are then stably inherited. Not surprisingly, the methyladenines present as a result of *dam* methylation at GATC sites (see Chapter 6.2) are completely lost following transfection of cloned DNA into animal cells. The same is true of all other methylcytosines except those in methyl CpG dinucleotides (Stein, Gruenbaum, Pollack, Razin & Cedar, 1982).

It is of course possible that a small proportion of the input DNA escapes being methylated by the HpaII methylase and that these molecules are selected for during subsequent growth and cell cloning procedures. To try to rule this out, Pollack *et al.* (1980) treated their *in vitro* methylated DNA with the restriction endonuclease HpaII to destroy any unmethylated molecules, but even before treatment they could detect only 0.001% of the DNA molecules in an unmethylated state.

Whether a transfected DNA molecule loses its methylcytosines or not may depend on its site of integration in the chromatin and/or on whether it is expressed. It may not be surprising, therefore, that a methylated thymidine kinase gene loses its methyl groups following transfection when it is present in a clone selected in HAT medium. (The next section looks at the *tk* gene in detail.) To allow for this, Stein, Razin, and Cedar (1982) cotransfected *tk⁻* cells with the HSV *tk* gene and a hamster adenosine phosphoribosyltransferase (*aprt*) gene and selected for TK⁺ clones. More than 95% of the clones arising by transfection with an unmethylated *aprt* gene expressed that gene, whereas almost all of the clones arising from an HpaII methylated *aprt* gene appeared APRT⁻. The few APRT⁺ clones arising from the methylated *aprt* DNA had undergone some demethylation.

It would appear then that, once established, a clone maintains its methylation pattern in the absence of selection. In the early stages following transfection, a marked amount of demethylation may occur, perhaps brought about by repair, recombination, and replication processes going on prior to stable integration.

On injection of double-stranded DNA into *Xenopus* oocytes no replication occurs, and no change occurs in the level of methylation (Vardimon *et al.*, 1982a). However, following injection into eggs, the methylated state at HpaII sites is maintained as the DNA replicates extrachromosomally (Harland, 1982). As the *Xenopus* egg develops, the injected DNA may become integrated with the result that mosaic frogs are obtained, some tissues containing the injected gene, others not. Some unintegrated episomal copies also persist (Andres *et al.*, 1984). There is evidence that injected globin genes maintain methylcytosines in four HpaII sites during replication and development (Bendig & Williams, 1983), but that in a plasmid containing the vitellogenin gene considerable demethylation occurs (Andres *et al.*, 1984). The reason for this difference is not clear, especially as the vitellogenin genes are not transcribed in young frogs nor can they be actived by estrogen treatment.

Although little *de novo* methylation occurs in somatic cells or in *Xenopus* eggs, the situation in early mammalian development is quite different (see Section 8.4.6).

8.4.2 Methylation and Expression of the Transfected Genes

8.4.2a The Herpes Simplex Virus Thymidine Kinase Gene (HSV *tk*)

When the HSV *tk* gene is transfected into TK$^-$ cells and transformants selected with HAT medium, it is found that the introduced gene remains unmethylated and is expressed so long as the cells remain in the selection medium. However, on removal of the cells to nonselective medium, there is an exponential decay in the proportion of cells expressing thymidine kinase. Similarly, on transfer of transformants to medium containing bromodeoxyuridine instead of HAT there is a counterselection for cells not expressing thymidine kinase, and *tk*$^-$ revertants arise at high frequency (about 10% of the cells revert to the TK$^-$ phenotype). This loss in expression is seldom caused by the loss of the HSV *tk* gene, and in most cases the gene remains largely unmethylated (at HpaII and HhaI sites) even though it is not being expressed (Pellicer *et al.*, 1980; Ostrander *et al.*, 1982; Davies *et al.*, 1982). Thymidine kinase can be reinduced in up to 50% of these nonexpressing transformants by transferring them back into HAT medium.

These changes in expression, which can be brought about at high frequency, are associated with corresponding changes in the nuclease sensitivity of the chromatin around the thymidine kinase gene. Such high frequency of change may be typical of certain regions of mammalian cell chromatin but it is not accompanied by methylation or demethylation of the DNA (Davies *et al.*, 1982; Roginski *et al.*, 1983). It affects not only the thymidine kinase gene whose expression is being selected for (or against) but also other genes linked to the HSV *tk* gene for transfection. This implies that the changes in chromatin structure associated with the changes in gene expression can take place over regions of DNA at least as long as 20 kbp.

In contrast to this high reversion frequency involving unmethylated DNA, a small number of the thymidine kinaseless cells contain an HSV *tk* gene that is heavily methylated (Pellicer *et al.*, 1980; Ostrander *et al.*, 1982; Clough *et al.*, 1982; Christy & Scangos, 1982). Such cells arise when transformants are grown in nonselective medium or in bromodeoxyuridine-containing medium. These cells no longer show high frequency switching when transferred to HAT medium and back to bromodeoxyuridine medium. One clone that reverted to a TK$^+$ phenotype in HAT medium was shown to have an unmethylated HSV *tk* gene. Treatment with 5-azacytidine increases up to 13-fold the frequency of reversion of these TK$^-$ in HAT medium, and again this treatment is associated with the loss of methyl groups from HpaII sites in the thymidine kinase gene. When the chromosomal DNA isolated from a cell containing a methylated HSV thymidine kinase gene was used to transfect a new *tk*$^-$ cell, no transformants were obtained, but DNA isolated from revertants of the same cell, which now expressed HSV thymidine kinase, produced several positive transformants per plate (Christy & Scangos, 1982).

In a later study, Christy and Scangos (1984) analyzed a greater number of TK-expressing or nonexpressing transformants. Most CpG sites analyzed failed to show a perfect inverse correlation between expression and methylation, but a particular 780-bp AvaI band was consistently present only in expressing cells. This band arose by cleavage of the *tk* gene at two sites spanning the 5′ end of the gene.

These results strongly suggest that once the HSV *tk* gene becomes methylated at certain sites it is not expressed. However, they cannot of themselves rule out the possibility that some change other than methylation is bringing about the inhibition of transcription as occurs in the high frequency switchers. Methylation may follow inhibition in some cases, and when this happens re-expression becomes a rare event. We must bear in mind the fact that this thymidine kinase gene is not normally methylated during the HSV lytic cycle, and hence *control* by methylation in this case is irrelevant. In cells transformed by HSV, the thymidine kinase gene is not usually active, but the location of methyl groups in this situation is not known. Two hexanucleotide sequences occur at about 50 and 90 bp 5′ to the HSV *tk* gene and are essential for its efficient transcription (McKnight *et al.*, 1984). Mutations in these sequences dramatically reduce transcription. As these sequences each contain one or more CpG dinucleotides, it would not be surprising if their methylation were to block transcription, and it would be important for the virus to have a means of replication in the absence of DNA methylation.

From this point of view, it is interesting that methylation using EcoRI methylase of an adenine in the sequence GAATTC, which occurs 79 bp upstream from the 5' end of the HSV *tk* gene, very markedly reduces expression when the gene is microinjectd into cells (Waechter & Baserga, 1982). This wholly unnatural methylation interferes with transcription, but its only relevance to the study of the control of transcription in eukaryotes is that it emphasizes the general assumption that alterations in base sequence in regulatory regions can have significant effects.

8.4.2b Expression of the Adenosine Phosphoribosyltransferase Gene (*aprt*)

Using the HSV *tk* gene as a selectable marker, Stein, Razin, and Cedar (1982) showed that on cotransfection with the adenine phosphoribosyltransferase (*aprt*) gene, the expression of the latter was blocked by prior methylation with the HpaII methylase. Revertants expressing *aprt* could be obtained at a frequency of up to 0.04% by growing the phenotypically APRT⁻ cells in azaserine, which inhibits purine biosynthesis. Reversion is associated with extensive demethylation and/or amplification of the *aprt* gene. Reversion could not be induced by treatment of cells with azacytidine.

8.4.2c Expression of Globin Genes

Using a similar selection technique, Chen and Nienhuis (1981) introduced human α and β globin genes into mouse fibroblasts. Although these genes remained unmethylated, only very low levels of globin transcripts were obtained. To some extent, this may be caused by lack of specific factors present in erythroid cells, which are required to transcribe the β globin genes (Wright *et al.*, 1984). Despite this, Busslinger *et al.* (1983b) and Busslinger and Flavell (1983) have shown that methylation of the γ globin and the α globin genes blocks their expression when they are introduced into fibroblasts along with the thymidine kinase gene. They did not use HpaII to methylate occasional CpGs, but used a technique described by Stein, Gruenbaum, Pollack, Rozin, and Cedar (1982) to methylate every cytosine in the input DNA.

Using globin DNA cloned into the single-stranded DNA phage M13, they constructed a complementary strand using a short synthetic primer, *E. coli* DNA polymerase I and four deoxyribonucleoside triphosphates: dATP, dGTP, dTTP, and 5-methyl dCTP (in place of dCTP). The resulting DNA molecules contain only methylcytosine residues in the newly synthesized strand (except for the primer), while the template strand remains unmodified, as shown in Figure 8.13. When introduced into vertebrate cells, the template strand of this hemimethylated molecule is methylated by the vertebrate cells' DNA methylase. Methylation then occurs exclusively and completely at CpG residues. Following replication, all CpGs

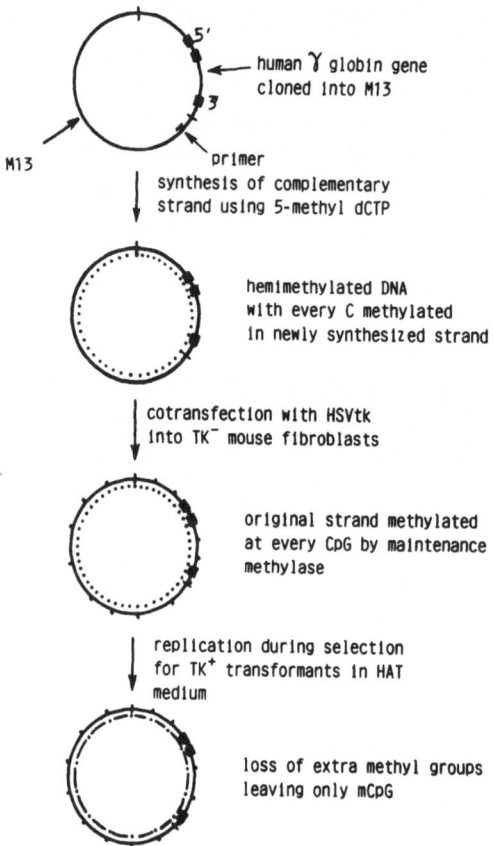

Figure 8.13. Method of preparing DNA with every CpG methylated. See text for details. Dots represent methylcytosines.

remain methylated, but the input DNA with methyl groups on all cytosines of one strand is gradually diluted out with successive divisions.

Cells transformed with DNA methylated throughout the γ globin gene and the M13 vector (except for a short primer) do not express the γ globin gene. In four different clones investigated, from five to 20 copies of the DNA were integrated and remained methylated to almost 100% at four HhaI and two HpaII sites in the globin gene. When cells were transformed with unmethylated DNA, it remained unmethylated and globin transcripts were synthesized.

By selecting primers covering different regions of the γ globin gene or M13 vector, Busslinger *et al.* (1983b) were able to look at the effects of methylating different segments of the DNA, as shown in Figure 8.14. When the whole γ-globin insert was used as primer, only the vector was methylated. When such DNA was introduced into mouse fibroblasts, the segmental methylation pattern was maintained and γ globin gene transcripts were synthesized in undiminished quantities showing that all the

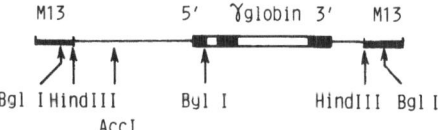

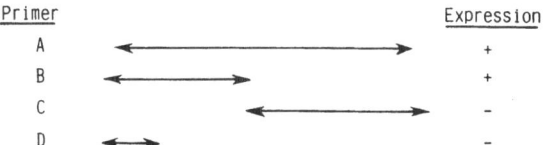

Figure 8.14. Effect of methylation on the expression of the human γ globin gene. The upper part of the figure represents the human γ globin gene and its flanking sequences cloned into M13. Duplex molecules were prepared as described in Figure 8.13, using the primers indicated in the lower part of the figure. All the rest of the molecule, except for that covered by the primer, becomes completely methylated. At the lower right is indicated the result of expression assays when these DNA's were introduced into mouse fibroblasts. (Busslinger, Hurst & Flavell, 1983b.) Copyright M.I.T.

extra methyl groups in the vector have no adverse effect on transcription of the cloned gene.

The γ globin insert can be cleaved into two parts with BglI. One fragment (C) stretches from a point just inside the gene (nucleotide + 100), throughout the gene, to the point just inside the vector. The second fragment (B) stretches from the point just inside the vector at the other side, through the 5′ flanking region, up to nucleotide + 100. When fragment B is annealed to the γ globin plasmid for use as a primer, the structural gene is methylated except for the first 100 bases. On its introduction into mouse fibroblasts, γ globin transcription occurs in the resulting clones. However, when fragment C is used as primer, the 5′ flanking region of the gene is methylated, and no γ globin transcripts are found in resultant clones. Using as primer a fragment stretching from −1350 to −760 produces cell clones containing a γ globin gene methylated from −760 in the 5′ flanking region and throughout the gene. Again, no γ globin gene transcripts are produced. The conclusion is that methylation within the region from −760 to +100 blocks expression of the γ globin gene. This region contains 11 CpGs, and we must await further experiments before we know which are important in blocking transcription.

8.4.2d Expression of Papovavirus DNA

When polyoma DNA is methylated with HpaII methylase, this has no effect on the infectivity of the DNA, and replication results in the loss of the added methyl groups (Subramanian, 1982). In the same way, SV40 DNA methylated at CCA/TGG sequences by cloning in *E. coli* readily infects cells, and the resultant viral DNA is fully sensitive to *EcoR*II, indicating loss of methyl groups (Adams, unpublished observations). However, methylation of SV40 DNA with HpaII methylase, which results in addition of only one pair of methyl groups per genome, does affect transcription of the late genes when this DNA is injected into *Xenopus* oocytes (Fradin *et al.*, 1982). Production of the large and small T antigens (the early gene products) is not diminished.

The HpaII site is 24 base pairs downstream from the major SV40 initiation site of late mRNA (there are also multiple minor late mRNA initiation sites). The promoter for SV40 late mRNA is unusual in that it lacks the typical TATA and CAAT boxes and is only active in permissive cells. Deletion studies have implied that transcription is activated by the passage of the replication fork (Contreras *et al.*, 1982), and hence the finding that transcription of late genes occurs at all in oocytes is unexpected. Nonetheless, transcription of late genes has also been obtained in cell-free systems, and here prior *in vitro* methylation with HpaII methylase has no effect (Manley *et al.*, 1980).

In the oocytes, the DNA does not replicate, but when SV40 or polyoma virus DNA is injected into tissue culture cells, it begins to replicate at about the same time as transcription of the late genes is initiated. Methylation of every CpG in SV40 or polyoma DNA using rat liver DNA methylase does not block expression of either early or late genes, nor does it reduce the efficiency of infection or oncogenic transformation of cells (Graessmann *et al.*, 1983). Obviously, the different effects of methylation of SV40 DNA on late gene expression can be explained by the assumption that replication of DNA has led to demethylation and hence has removed the obstacle to transcription. Indeed, analysis of DNA 40 hours after injection or in mature virions shows that it is now unmethylated. In three clones of oncogenically transformed cells investigated, the DNA had become partially or completely demethylated at the HpaII site.

The use of the eukaryotic DNA methylase in this type of experiment is to be strongly commended. It does not suffer from the disadvantage of the HpaII methylase in that it methylates every CpG. Nor does it introduce the potential artifact of methylating every cytosine, as is the case for the method of Stein, Razin, and Cedar (1982; Figure 8.13).

8.4.2e   Expression of Retrovirus Proviral DNA

The contrast between the use of the rat liver methylase and the HpaII methylase is shown by their effect on the biological activity of Moloney murine leukemia virus (M-MuLV) proviral DNA (Simon *et al.*, 1983).

Although methylation of 35 HpaII sites had no effect on infectivity when the proviral DNA was transfected into mouse fibroblasts methylation of every CpG with the rat liver enzyme reduced infectivity by at least three orders of magnitude. Moreover, in cells transfected with *in vitro* methylated proviral DNA, the dormant DNA can be reactivated by treatment with azacytidine.

When the heavily methylated endogenous M-MuLV proviral DNA is transfected into mouse fibroblasts, it shows low infectivity. In contrast, exogenous, hypomethylated, proviral DNA from the same tissue is highly infectious (Stuhlmann *et al.*, 1981). In this case, we are comparing the same DNA in its naturally methylated and unmethylated forms (see Section 8.3.3b). Similar results have been found for murine leukemia proviral DNA by Hoffman *et al.* (1982), but cloned Moloney sarcoma proviral DNA will not transform cells if it is previously methylated with HpaII and/or HhaI methylases (McGeady *et al.*, 1983). By cotransfection with plasmids containing different regions of the MSV provirus, it was shown that methylation of the v-*mos* gene is more inhibitory to transformation than methylation of the viral LTR (see also Section 8.3.3b).

8.4.2f Expression of Adenovirus DNA

Cloned, unmethylated DNA from the E2a region of adenovirus 2 (see Figure 8.5) is readily transcribed on injection into *Xenopus* oocytes (Vardimon *et al.*, 1982a). Prior *in vitro* metylation with the HpaII methylase (but not with the BsuRI methylase, which methylates the internal cytosine in the sequence GGCC) abolishes transcription from the viral promotor. There are 14 HpaII sites in the cloned E2a region, but methylation of the 11 sites in the main part of the gene region does not prevent correct initiation of Ad 2-specific transcripts. However, methylation of the three upstream CCGG sequences (at -215, $+5$, and $+23$ bp) completely blocks transcription in *Xenopus* oocytes (Langner *et al.*, 1984).

When the promoter for the adenovirus type 12 early protein (Ela) gene is linked to the chloramphenicol acetyltransferase gene (*cat*) and transfected into mouse cells, the enzyme CAT is synthesized whether or not the *cat* gene is previously methylated at four internal HpaII sites. The Ela promoter (the 525 bp at the left end of the genome, see Figure 8.5) carries two HpaII and three HhaI sequences, and *in vitro* methylation of either of these abolishes the expression of the linked *cat* gene. If the Ela promoter is replaced by the promoter for the adenovirus protein IX gene, which carries four HpaII or HhaI sites, *in vitro* methylation of these has no effect on transcription (Kruczek & Doerfler, 1983). This illustrates the disadvantage of using the bacterial methylases; one cannot conclude that methylation is not important when it has no effect. It is very likely that the important sites have not been methylated.

8.4.2g   Comment

These observations illustrate two points. First, methylation of some sites, usually toward the 5' ends of genes, can block transcription. Not all CpGs are important in this respect, and so this is not an explanation for the presence of the majority of the methylcytosines in the cell. Second, because *in vitro* modifications can block transcription, one must not immediately conclude that we are studying a physiological control mechanism. We must obtain evidence that such controls actually apply in normal circumstances. Thus, although methylation of the region upstream of the human γ globin gene blocks its expression, we do not know that the vital CpGs are ever methylated *in vivo*. If not, then methylation plays no direct part in control of gene expression, and instead we must seek a mechanism that ensures that particular sites are never methylated. Failure of such a mechanism would, in effect, lead to an heritable change in base sequence and would act as a mutation.

8.4.3   The Question of *de Novo* Methylation in Early Mouse Embryos and Teratocarcinoma Cells

A mammalian egg is a thousandth or less of the volume of an amphibian egg, and it is only recently that techniques have been developed for injecting DNA into such small cells. Moreover, obtaining and culturing large numbers of mammalian eggs is far more of a problem than working with *Xenopus* eggs. The necessary techniques were developed to help investigate the totipotency of somatic cell DNA. Nuclear transplant experiments in mammals, where the pronuclei of the recipient egg are removed and replaced with another nucleus, have been unsuccessful in achieving development beyond cleavage. The only example of development to term occurs when the donor nucleus comes from the inner cell mass cells of blastocysts. Nuclei from trophectoderm cells from similar blastocysts only support development to midcleavage (Illmensee & Hoppe, 1981). Similar results using a gentler technique (McGrath & Solter, 1983) also imply a rapid loss of totipotency very early in development. Such a loss might conceivably result from the irreversible establishment of a particular pattern of methylation, which helps to direct the pathway of differentiation. It is interesting to note that the DNA of the inner cell mass cells retains the higher level of methylation found in the egg at a time when the DNA of trophectoderm cells has already fallen to the level of methylation found in most somatic cells (see Section 8.2). We should not, however, lose sight of other possibilities, such as DNA rearrangements, or simply the establishment of certain stable chromatin structures, as a cause of the apparent, early loss of totipotency.

8.4.3a   Retrovirus Proviral DNA

As reported above (Section 8.5.5b), the infectivity of Moloney mouse leukemia virus (M-MuLV) proviral DNA on transfection into mouse *fibroblasts* depends on its state of methylation. However, when unmethylated proviral DNA is microinjected into mouse *zygotes* it is not expressed, but rather it is integrated and rapidly becomes methylated to between 70 and 95% at 44 HhaI, AvaI, and HpaII sites tested (Jahner *et al.*, 1982). A similar result is found when 4 to 16 cell pre-implantation embryos are infected with M-MuLV; but on infection of post-implantation, 8-day embryos (10^4–10^5 cells) with purified virus, multiple foci of cells expressing viral RNA are found in a variety of tissues on analysis four days later. The proviral DNA isolated from liver, kidney, and brain cells is methylated to less than 20%.

Proviral DNA integrated into pre-implantation embryo chromosomes may become activated later in development. Whether or not it does depends on the chromosomal location of the provirus and the tissue involved (Jaenisch *et al.*, 1981; Jaenisch & Jahner, 1984). Thus in one mouse substrain (Mov-1) carrying a MuLV genome experimentally inserted into the germ line of the original recipient, no spontaneous activation occurs in fibroblasts or hematopoietic cells; whereas, in a second substrain expression occurs only in muscle cells (Fiedler *et al.*, Harbers *et al.*, 1981). This suggests that transcriptional competence of the chromosomal region surrounding the site of provirus integration determines whether or not viral expression will occur.

De novo methylation also occurs when M-MuLV proviral DNA is transfected into murine teratocarcinoma (embryonal carcinoma or EC) cells (Gautsch & Wilson, 1983; Linney *et al.*, 1984; Niwa *et al.*, 1983; Jahner *et al.*, 1982; Stewart *et al.*, 1982). These are the malignant and pluripotential stem cells derived from carcinoma of pre-implantation stage embryos. Thus the EC cells are a convenient model for pre-implantation embryos.

However, in this case the methylation does not occur for several days after infection, yet little or no viral gene transcription is detected even prior to methylation, implying that mechanisms other than methylation initially block expression. Supporting this conclusion, Niwa *et al.* (1983) showed that suppression of DNA methylation with 5-azacytidine does not result in activation of the viral genome.

Fusion of infected EC cells with differentiated cells results in virus production. This result can be explained if a necessary function is lacking in the EC cell or if a specific inhibitor is present (Gautsch, 1980). Linney *et al.* (1984) have shown that the primary block in M-MuLV infection is the inability of a pair of tandem repeats that normally function as enhancer sequences to function in EC cells.

The EC cells can be induced to differentiate by a number of techniques, and only in infected, differentiated cells is the retrovirus proviral DNA activated by treatment with 5-azacytidine (Niwa *et al.*, 1983).

Thus it would appear that the block to transcription in EC cells and probably in pre-implantation embryos is the lack of a necessary factor. The inactive gene is then subject to *de novo* methylation, which occurs relatively slowly in EC cells. The methylated genes may be activated later in a tissue-specific manner in differentiated cells, and this activation can also be brought about by azacytidine treatment, though whether the *primary* action of the drug is to bring about demethylation or not is uncertain (see Section 5.4.3).

8.4.3b   Papovavirus DNA

As with retroviruses, polyoma virus will not normally grow in mouse embryonal carcinoma cells, although other mouse cells are the usual host for this virus. Several polyoma mutants that will grow in undifferentiated EC cells have been obtained (Fujimura *et al.*, 1981; Sekikawa & Levine, 1981; Herbomel *et al.*, 1981). These mutants all show mutations in a small region around the origin of replication and the initiation site of late transcription. Although little change in chromatin structure (as assayed by DNase I sensitivity of cellular minichromosomes) is found, many mutants have a short duplication and contain a G for A base substitution in this region.

Although polyoma DNA may become methylated in EC cells (and there is no evidence that it does), there are no CpG dinucleotides in the area affected by the facilitating mutations. One must therefore conclude that again in this case the growth restriction is unrelated to DNA methylation.

By injecting a plasmid containing the early region of SV40 DNA into the male pronucleus of fertilized single-cell eggs, transgenic mice were obtained in which every cell carried the SV40 sequences (Brinster *et al.*, 1984). Almost all of these animals developed brain tumors. In the normal cells of these animals the SV40 DNA was methylated at HpaII, SmaI, and XhoI sites, but in the tumor cells most of these sites were largely unmethylated. The implication of these studies is that the SV40 genes are inactivated early in development and reactivated later in a tissue-specific manner. The active genes are undermethylated.

8.4.3c   Immunoglobulin and Metallothionein Genes in Transgenic Mice

When rearranged immunoglobulin genes or a fusion plasmid containing the mouse metallothionein-I promoter fused to the HSV *tk* structural gene are injected into fertilized eggs, the eggs can be reinserted into foster mothers, and will develop into adult transgenic mice. These transgenic mice can be mated, and the expression of the transferred genes can then be studied in their offspring. The immunoglobulin κ gene is only expressed in spleen and not in liver cells (Brinster *et al.,* 1983). Mice carrying the metallothionein/thymidine kinase gene have activity that can be induced in the liver by heavy metals, but no induction was found with glucocorticoids (Palmiter *et al.,* 1982, 1984). In some, but not all, mice the DNA becomes methylated after injection; but when tested in adult mice, inserts in different locations appear differentially methylated, again emphasizing the importance of chromatin structure in determining methylation patterns. There is loss of methylated sequences in some cases from one generation to the next, and only a poor inverse correlation with gene expression.

8.4.3d   Conclusion

One can conclude that in early embryos there is a block to the expression of the genes studied and that *de novo* DNA methylation probably follows inactivation. Early methylation does not prevent, and may be essential for, later expression in a tissue-specific manner.

9

X-Chromosome Inactivation

9.1 The Phenomenon

Female mammals have two X-chromosomes, but male mammals have only one X-chromosome. This could result in females having twice as many X-chromosome transcripts as males, but this does not occur because one of the X-chromosomes is largely inactive. (The tip of the short arm of both X-chromosomes is active.) The inactive X-chromosome is highly condensed, replicates late in S phage, and can be seen as the dense Barr body in interphase nuclei. The active region of the inactive X-chromosome is not late replicating and can be detected in chromosome spreads by its increased sensitivity to nucleases (Kerem *et al.*, 1983). In XXY males with an extra X-chromosome (Klinefelter's syndrome), one X-chromosome is inactivated.

In general, the inactive X-chromosome can be derived from either the paternal or maternal germ cell. In eutherian mammals, two exceptions to randomness are known: the extra-embryonic trophectoderm (this is a unipotential tissue that does not form part of the developing embryo), and primary endoderm cells of the early embryo in which the paternal X-chromosome is preferentially inactivated. (Marsupials preferentially inactivate the paternal X-chromosome in all cells.)

The X-chromosome contributed by the sperm is inactive at the time of fertilization but is reactivated by the 8 to 16 cell stage, prior to the random inactivation that occurs later. In females, reactivation of the inactive X-chromosome occurs at the time the female germ cells enter meiotic prophase and both X-chromosomes function throughout oogenesis.

As cells of the embryo differentiate, X-inactivation occurs. First the cells of the trophectoderm and shortly afterwards cells of the primary endoderm differentiate, but as these show preferential inactivation of the paternally derived X-chromosome, it is not certain that this chromosome ever becomes fully activated in these cells.

By 6.5 days of development the majority of cells in the mouse embryo have only one active X-chromosome (Monk & Harper, 1979; Martin, 1982). Thereafter, the inactive X-chromosome is inherited from cell generation to generation such that every female is a mosaic containing clonal groups of cells expressing genes carried on either the maternally or paternally derived X-chromosome. As clonal groups of cells tend to stay close together, this results in clusters of cells expressing one or the other X-chromosome (Lyon, 1972).

Occasionally, a part of an autosome is translocated onto an X-chromosome. When X-chromosome inactivation occurs, the chromatin condensation spreads into the adjacent autosomal region (Cattanach, 1975). However, for inactivation to occur, a region of the X-chromosome known as the primary X-inactivation center is necessary, and it is proposed that the decision as to which X-chromosome is to be inactivated affects this site. The inactivation then spreads to the rest of the chromosome.

9.2   Two Models to Explain X-Chromosome Inactivation

As either of the X-chromosomes can be inactivated, and as the chromosome environment must be virtually identical in two cells, one of which inactivates the paternal X-chromosome and one the maternal X-chromosome, the nucleotide sequence *per se* can have no role to play in inactivation. The choice as to which chromosome is inactivated is random, and there is no evidence concerning what is the initial inactivating event, although a number of models have been proposed (see Riggs, 1975).

There are two suggestions as to how inactivation may be inherited, once established. The chromatin structure itself may be inherited, and the cooperative alignment of chromosomal proteins could lead to pattern inheritance, shown in Figure 9.1. The alternative suggestion (and the two are not mutually exclusive) is that the inactive X-chromosome is heavily methylated, and because of the way methylation patterns are inherited (see Section 2.3), this leads to the production of clones of cells with either the maternal or paternal X-chromosome inactivated. We will look at the evidence for this in the following section, but it should be made clear that the features of the DNA methylation model that make it attractive for the faithful copying of methylation patterns make it less attractive as an explanation for the spreading of inactivation along the chromosome. Thus, if the methylase acts processively from an initial inactivation point (see Section 5.2), then it may lead to the spread of methyl groups along the chromosome. But a similar blanket methylation would be expected for all chromosomes, as none of them are free of methyl groups. The inactivation event would involve *de novo* methylation which, as reported in Sec-

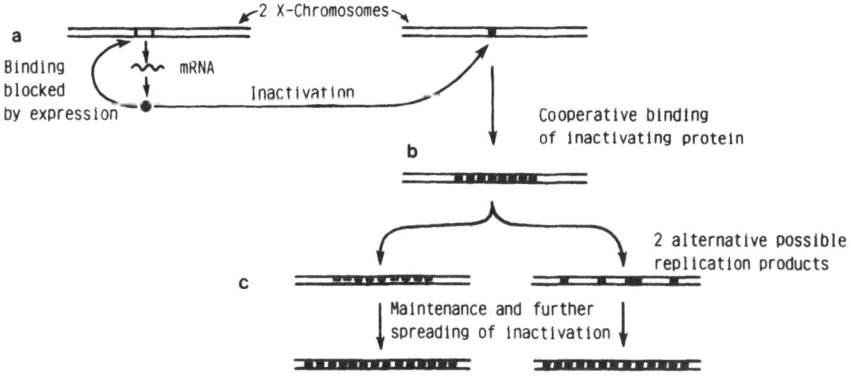

Figure 9.1. X-chromosome inactivation and chromatin proteins—a suggestion.
(a) Production of inactivator protein, which binds to its own promoter unless that
promoter is active, *i.e.*, binds only in *trans*. (b) Cooperative binding of inactivator
protein (or some other protein seeded by inactivator protein) inactivates X-chromosome. (c) At replication, inactivator protein distributed to daughter chromosomes (two possible modes of distribution are indicated). Cooperative binding
leads to continued inactivation and further spreading of inactivation.

tion 8.4.6, only occurs to a significant extent in pre-implantation embryos.
As this is the time when inactivation occurs, this could provide supportive
evidence that methylation does play an important role in inactivation.
However, it may be that at this stage in development any inactive DNA is
methylated and thereby ticketed for future cell generations.

The alternative model, which relies on the cooperative binding of chromosomal proteins, could similarly rely on the absence of the inactivating
protein from other chromosomes, *i.e.*, the presence of a specific inactive
X-chromosome-associated protein or modified protein. But no such protein has yet been identified.

It is not the absence of a Y-chromosome but the presence of two or
more X-chromosomes that leads to inactivation of all but one X-chromosome. Obviously, if the single X-chromosome in a normal male cell is
inactivated, that cell will cease to function, but there is no evidence for
the wholesale cell death that would be expected. Whatever initiates inactivation, it may be an *one off* event, such as the binding of an X-chromosome to a single intranuclear site, leading to the inactivation of all un-
bound chromosomes.

Alternatively, a control loop may be established such that the first X-
chromosome to express an *inactivation protein* leads to inactivation of all
other X-chromosomes and continued activity of the first X-chromosome
(see Figure 9.1).

The selection, propagation, and maintenance aspects of inactivation
may each involve a separate mechanism. The evidence presented below
provides strong support for a role of DNA methylation in only the maintenance of the inactive state.

Embryonal carcinoma (EC) cells from female mammals have both X-chromosomes active as long as the cells are maintained in an undifferentiated state. Fusion of EC cells with thymocytes or bone marrow cells from female mice leads to the reactivation within four days of the inactive X-chromosome from the somatic cell (Takagi *et al.*, 1983). It is difficult to reconcile this finding with a role for *de novo* methylation in X-chromosome *inactivation*. Rather it implies the presence of a factor in the EC cells that reverses the inactivation process. Similar activation of inactive genes often occurs on cell fusion—thus the whole inactive hen reticulocyte nucleus is reactivated on fusion with HeLa cells (Harris, 1970), and human muscle proteins are made within 24 hours of fusion of human nonmuscle cells with mouse muscle cells (Blau *et al.*, 1983).

9.3   Differential Levels of Methylcytosine in X-Chromosomes

The evidence presented in the following section reveals a difference in the DNA from the active and inactive X-chromosomes, and shows that, under certain circumstances, genes on the inactive X-chromosome can be reactivated by azacytidine treatment. However, 5-methylcytosine-specific antibodies fail to discriminate between active and inactive X-chromosomes (Miller *et al.*, 1982), and, using probes to X-chromosome-specific DNA sequences selected at random, researchers have found no evidence for differential methylation of HpaII sites (Wolf & Migeon, 1982, 1983). Instead, the pattern of X-chromosome methylation in normal human cells tends to vary from cell to cell. The extent of methylation, as measured by the size of the HpaII fragments hybridizing to an X-chromosome-specific probe, varies between clones of male cells in all of which the single X-chromosome is active. However, when a similar analysis is performed on DNA from female cells, no new large HpaII fragments are found, indicating that the region probed is not more highly methylated in the inactive X-chromosome.

The nature of the DNA recognized by these random probes is unknown, and one must realize that on the X-chromosome, as on all chromosomes, there are loci that may be hormonally controlled, developmentally regulated, or constitutively expressed; and each will be made up of exons and introns and separated by spacer DNA. The significance of the above results is, therefore, not clear.

Of greater interest are the results relating to the level of methylation of the X-chromosomal gene for hypoxanthine phosphoribosyltransferase (HPRT). This is a *housekeeping gene,* spanning more than 30 kbp and containing nine exons, which is active in all tissues studied, including sperm (Venolia & Gartler, 1983b). Using several probes to sequences around the first intron and the third exon, Yen *et al.* (1984) looked at the level of methylation of 14 sites of the methylation-sensitive restriction enzymes HpaII (CCGG), HhaI (GCGC), AvaI (CPyCGPuG), and SmaI (CCCGGG). Wolf *et al.* (1984a) made a similar study using a probe to the first intervening sequence. A striking feature of this gene (and several others, see Section 7.2.3) is the presence of a cluster of CpGs at the 5' end that makes it difficult to define precisely which one or ones are unmethylated.

A comparison of DNA from males and females indicates that in the active chromosome, sites near the 5' end of the gene are unmethylated while all distal sites are methylated. The correlation between expression and hypomethylation of the 5' end of the gene is not perfect, however, as some sites are also unmethylated in DNA from the inactive X-chromosome. Indeed, Wolf *et al.* (1984a) show that the inactive allele can have a variety of methylation patterns. Also, at least one site is consistently undermethylated in the middle region of inactive genes.

Note that the cluster of CpGs at the 5' end of the *hprt* gene are, however, largely methylated in the inactive X-chromosome.

Mouse–human cell hybrids carrying only the inactive human X-chromosome also have an *hprt* gene more heavily methylated at the 5' end than those hybrids carrying an active human X-chromosome. Reactivation by treatment with azacytidine (see Section 9.3) is again associated with decreased methylation, but this was not necessarily spread throughout the gene. In some reactivated clones, methylated sites are intermixed with unmethylated sites, even near the 5' exon.

Another X-linked, housekeeping gene is that specifying glucose 6-phosphate dehydrogenase. This *g6pd* gene has a cluster of CG dinucleotides at its 3' end and in active chromosomes all but nine of these are extensively methylated. Of the nine sites unmethylated in DNA from male lymphocytes two or more are methylated in DNA inactive X-chromosomes from female lymphocytes. Although expressed in all cells the frequency of transcription of the *g6pd* gene varies between tissues and certain CG dinucleotides present around the transcription termination site show a *positive* correlation between methylation and expression. (Toniolo *et al.*, 1984; Wolf & Migeon, 1984; Wolf *et al.*, 1984b; Wolf & Migeon, 1985; Battistuzzi *et al.*, 1985; Mondello & Goodfellow, 1985).

Direct activation of a gene may involve an active process that requires or produces a definitive methylation pattern. When the gene is not active, the methylation pattern may be allowed to drift, and gradually the gene undergoes *de novo* methylation leading to the patterns seen on inactive X-chromosomes. Such drifting via *de novo* methylation is a characteristic of clones of fibroblasts and lymphoblasts (Wolf *et al.*, 1984a; Jolly *et al.*, 1983).

9.4   Gene Transfer Experiments

The strongest evidence that X-chromosome inactivation involves a change in the DNA is the finding that DNA from the inactive X-chromosome is inactive in DNA-mediated gene transfer experiments, while DNA from the active X-chromosome will transform the same cells.

The gene for hypoxanthine phosphoribosyltransferase (*hprt*) is on the X-chromosome and is defective in humans suffering from Lesch-Nyhan syndrome. Thus cells cloned from Lesch-Nyhan heterozygotes will have either a functional *hprt* gene on the active chromosome and will be capable of growth in HAT medium, or the functional gene will be inactive and the cells will not grow on HAT medium. DNA purified from two such clones was used to transfect HPRT⁻ mouse fibroblasts. Conflicting results have been obtained for this experiment. In one case the two DNA's yielded essentially the same number of transformants (de Jonge *et al.*, 1982), whereas in a second set of experiments the DNA from the active X-chromosome gave transformants at more than 20-fold the frequency of DNA from the inactive X-chromosome (Venolia & Gartler, 1983a). There were a number of differences in the experimental procedures used by the two groups, and possible reasons for the different findings are discussed in the more recent paper.

When a similar experiment was performed in which the donor and recipient cells were both from rodents, the DNA from the inactive X-chromosome was ineffectual and produced no transformants (Liskay & Evans, 1980). Similar experiments by Chapman *et al.* (1982) and Kratzer *et al.* (1983) have confirmed that inactive X-chromosome DNA from several tissues of *adult* female mice is inefficient in gene transfer experiments, but that DNA from the inactive X-chromosome present in embryonic yolk sac endoderm is functional in gene transfer. As mentioned above, X-chromosome inactivation occurs earlier and in a nonrandom manner in primitive endoderm cells, and the results of Kratzer *et al.* (1983) may indicate that in this tissue, inactivation is not associated with changes (modification) of the DNA.

These experiments, in general, indicate that the DNA in the inactive X-chromosome differs from that of the active X-chromosome. No chromosomal proteins are transferred along with the DNA, and as the primary sequence of the DNA is assumed to be the same in both cases, the implication is that the DNA from the two sources is differentially methylated, at least in certain key regions.

9.5 Reactivation with Azacytidine

Human–mouse cell hybrids that contain only the inactive human X-chromosome and no mouse HPRT are unable to grow in HAT medium. Treatment of such hybrid cells with 5-azacytidine at 2 to 5 μM for 24 hours, followed by selection in HAT medium, results in a 1000-fold increase in the number of positive clones over untreated cells. The whole of the X-chromosome is not reactivated, however, and it remains heterochromatic (Mohandas *et al.*, 1981; Jones *et al.*, 1982). A significant amount of cell death occurs upon treatment with azacytidine, but between 0.1 and 8% of the surviving cells form colonies in selective medium. This is a remarkably high proportion in view of the very limited activation of other genes on the X-chromosome, and it may imply some special sensitivity of the *hprt* gene. Indeed, Wolf and Migeon (1982, 1983) were totally unable to obtain activation of some other regions of the X-chromosome.

The dose of azacytidine that gives maximum activation of the inactive *hprt* gene is not the same as that giving maximum demethylation, and Shapiro and Mohandas (1983) concluded that although demethylation may facilitate reactivation, it does not induce it. Once again, it may be the case that demethylation is necessary, but not sufficient to induce gene expression. It is possible that incorporation of 5-azacytosine into DNA causes changes in chromatin structure over a limited area of the inactive X-chromosome and also brings about a general demethylation. Where the two coincide (as in the *hprt* gene), local activation may occur.

The finding that the *hprt* gene on the inactive X-chromosome is less efficient in a transfection assay is what would be expected if (1) this gene were highly methylated, and (2) methylation blocked expression. The findings are therefore consistent with a function of DNA methylation in the control of transcription. Moreover, in this instance, the alterations occurring to the DNA on X-inactivation are occurring within the cell under normal physiological conditions and so cannot be described as artificial in any way. Yet it is quite possible that, following chromosome inactivation, methyl groups are introduced into DNA in such a way that they block expression when this DNA is introduced into cells, but that such a methylation event is not normally used to control gene expression. It would be of considerable interest to know to what extent the inhibitory effect is found for other genes on the X-chromosome, and if it also applies to the *Xg* and *Sts* loci, which are on the active tip of the short arm of the otherwise inactive X-chromosome. Failure to detect gross differences in the level of methylation of active and inactive X-chromosomal DNA is perhaps not surprising. If the correlation is between *transcription* and undermethylation, one would expect that only about 10% of the DNA of even an active X-chromosome is transcriptionally active. This assumes that the X-chromosome is typical of the total cellular chromosomes. Furthermore, if the inverse correlation between transcription and methylation only holds for a few sites at the 5' end of a gene, then the overall DNA methylation level of active X-chromosomes would be expected to be about 99% of that found in inactive X-chromosomes.

10

DNA Methylation and Cancer

It goes without saying that if changes in the methylation of particular sites in DNA can cause changes in gene expression, then aberrant changes may lead to aberrant patterns of expression and possibly to cancer. What we shall review in this chapter is whether treatment of cells with carcinogens can lead to changes in DNA methylation, and whether or not tumor cells are characterized by higher or lower levels of methylcytosine in their DNA.

10.1 DNA Damage and Methylation

When DNA is damaged it is quickly repaired by excision or by post-replication repair mechanisms (see Chapter 6). Treatment of cells with nitrogen mustard or ultraviolet light or even the unbalanced growth that is involved in the technique of chemical synchronization of cells leads to damage and repair of DNA (Drahovsky *et al.*, 1976; Hilliard & Sneider, 1975; Lieberman *et al.*, 1983). The DNA made during the repair reaction is quickly methylated, but whether the extent of methylation is the same as that present initially is not clear. Using incorporation of radioisotopes into methylcytosine and other DNA bases as a measure, there appears to be, if anything, an increased level of methylation in the repair patches, but following ultraviolet irradiation at least one allele of the metallothionein-I gene, for instance, becomes largely demethylated. If this latter finding occurred generally throughout the genome, the level of demethylation should be very obvious, and hence one must conclude that in the case of the metallothionein gene some selection for unmethylated sequences may have occurred during the clonal growth of cells necessary for DNA analysis.

The decrease in methylcytosine content of bovine and rat DNA with increased age of the animal (Romanov & Vanyushin, 1980; Vanyushin *et al.*, 1973a) may be the result of the failure to methylate fully patches of repair. With the years, such undermethylation will become increasingly apparent and increasingly likely to occur in critical regions and may well contribute to the aging process, especially in more long-lived animals. In rats the changes with age are different in different tissues, possibly representing the balance between DNA repair and cell replacement, which can occur. Thus brain DNA loses most methylcytosine with age (a 14% drop), yet little change occurs in liver DNA, where regeneration is possible. On the other hand, this explanation does not account for the fact that spleen DNA loses 17% of its methylcytosine, and kidney DNA actually shows a 30% increase in methylation.

The undermethylation in older animals is found largely in the moderately and highly repeated sequences where it can have little effect on coding potential, but where it might lead to an elevated level of genome rearrangements (see Section 7.3.2).

10.2 Carcinogens and DNA Methylation

Alkylating agents such as N-nitro-N-methylurea and N-nitroso-N-methyl nitroguanidine (NMNG) are carcinogenic, and, as pointed out in Section 5.4.2, they interfere with DNA methylation *in vitro* by interacting both with the DNA and the enzyme. Treatment of human Raji cells with these reagents also leads to a reduction in the level of DNA methylcytosine (Boehm & Drahovsky, 1981a, 1981b, 1983).

Fluorene derivatives such as N-acetylaminofluorene are highly carcinogenic and interact with DNA to form an adduct on the C8 of guanine bases. Such DNA is poorly methylated both *in vitro* and in *in vivo* (Ruchirawat *et al.*, 1984; Pfohl-Leszkowicz *et al.*, 1981; Lapeyre & Becker, 1979; Lapeyre *et al.*, 1981).

Azacytidine is not usually regarded as being carcinogenic and is in fact used to treat leukemia (Wilson *et al.*, 1983). Yet this drug does cause a marked reduction in the level of DNA methylation, and hence the finding that carcinogens reduce the level of DNA methylation may be unrelated to their tumorigenic action. However 5-azacytidine and 5-fluorodeoxycytidine do induce chromosomal aberrations and can induce tumorigenic virus particles in transformed hamster cell lines (Chambers & Taylor, 1982; Altanerova & Altaner, 1972), and increase the rate of spontaneous leukemia in AKR mice (Vesely & Cihak, 1973). Azacytidine can also alter the metastatic phenotype of mouse lung carcinomas (Olsson & Forchhammer, 1984). Stable diploid hamster fibroblasts demonstrate tumor-forming ability after a brief treatment with azacytidine, and this transformation is associated with trisomy for chromosome $3q$ and demethylation at specific CCGG sites tested (Harrison *et al.*, 1983). It is speculated that oncogenes (Ha-*ras* and Ki-*ras*) and possibly the preproinsulin gene on chromosome 3 are activated on the extra copy, which arises as a result of azacytidine treatment.

It is clear then that certain carcinogens do bring about a reduction in methylation of DNA, but there is no direct relationship between carcinogenicity and the ability to interfere with methylation. Nevertheless, simply feeding rats a diet deficient in methyl groups induces hepatic carcinomas and enhances the carcinogenic activity of diethylnitrosamine. Such a methyl-deficient diet does lead to the production of DNA with a reduced methylcytosine content, but whether this is the cause of the carcinomas has yet to be established (Mikol *et al.*, 1983; Wilson *et al.*, 1984).

10.3   Changes in DNA Methylation Associated with Tumors and Transformed Cells

The analysis by Gama-Sosa *et al.* (1983b) of the methylcytosine content of 103 human tumors, which is diagrammed in Figure 10.1, illustrates the wide variation found. On comparing the benign tumors with normal tissues, one finds no difference in the average methycytosine content. Thus the DNA from normal tissues contains, on average, 0.905 mol % methylcytosine (or 4.37% of cytosines are methylated), and the DNA from benign tumors has 0.895 mol % methylcytosine (4.32% of cytosines are methylated). DNA from primary malignant tumors has on average 0.835

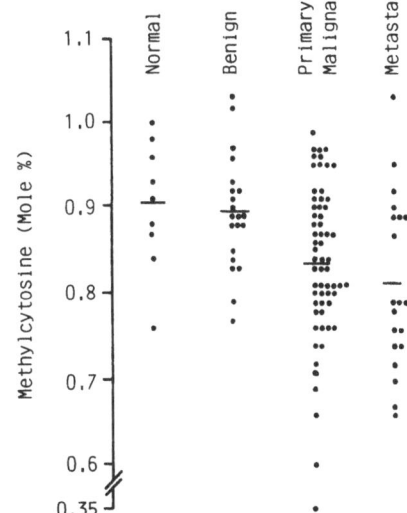

Figure 10.1. Methylcytosine content of the DNA from 103 human tumors and normal cells. Each point represents the determination on a particular cell or tissue, and the short lines indicate the average value for each column. The information is from Gama-Sosa *et al.*, 1983b.

mol % methylcyosine (4.06% cytosines methylated) and from metastases, 0.81 mol % methylcytosine (3.92% cytosine methylated). Although the figures from the malignant tumors are lower on average, this may be fortuitous as the range of values is greater than that for normal tissues.

Bearing in mind the spread of results presented in this survey, it is not surprising that on looking at individual tumors or transformed cells, different workers report no change (Diala & Hoffman, 1982), enhanced (Marcaud et al., 1981; Eick et al., 1983; Gunthert et al., 1976; Desai et al., 1971; Rubery & Newton, 1973; Cato et al., 1978), or reduced (Smith et al., 1982; Feinberg & Vogelstein, 1983a, 1983b; Kuhlmann & Doerfler, 1982; Gattoni et al., 1982; Lapeyre & Becker, 1979; Romanov & Vanyushin, 1980; Reilly et al., 1982; Ehrlich et al., 1982) levels of methylation. In some cases when individual genes are examined, not only does the pattern of methylation vary when cells are grown from the tumor, but the number of copies of the transforming virus also varies, making any correlation between undermethylation and transformation difficult to sustain. (e.g., Kuhlmann & Doerfler, 1982). The variable undermethylation found in the DNA of certain mouse hepatic tumors is not a direct result of a lowering in the amount of DNA methylase present. Indeed, not surprisingly, it was found that the tumors have an elevated level of enzyme relative to the nondividing control liver tissue (Lapeyre et al., 1981).

Although viral oncogenes are generally undermethylated in transformed cells, the cellular counterpart may be fully methylated or undermethylated (Gattoni et al., 1982; Feinberg & Vogelstein, 1983a, 1983b). In primary human adenocarcinomas, the cellular oncogenes c-Ha-ras and c-Ki-ras are undermethylated relative to adjacent sequences, but the same is true for the growth hormone gene and the γ globin gene, and no difference was seen for the α globin gene. Perhaps confirming the result of Gama-Sosa et al. (1983b), a metastasis from one patient showed an even greater degree of hypomethylation than did the primary tumor.

It is very tempting to generalize from the metastatic tumors with a reduced level of methylation that there is an inverse correlation between DNA methylation and malignancy. There are, however many exceptions and no evidence as to whether changes in DNA methylation are a cause or a consequence of oncogenic transformation. Certainly the expression of some oncogenic viruses is blocked by prior methylation, but this is an effect on a specific gene rather than on cellular DNA in general. Excessive repair of DNA following exposure to carcinogens or ultraviolet light may lead to variations in the extent of DNA methylation. When this occurs at particular sensitive sites it could lead to changes in gene expression, which may lead to cell transformation. As yet, however, this is largely speculation.

11

Variations on a Theme: Patterns of Methylation in Protista, Fungi, Plants, and Animals

11.1 Introduction

In previous chapters, discussion has focused on the extensive methylation of cytosines adjacent to guanine residues in the DNA's of vertebrates and echinoderms. It will be recalled that besides considerable variation in DNA methylation patterns between different species and even between the tissues of individual organisms, two general themes were discernible, the so-called *vertebrate-pattern* and the *echinoderm-pattern* (see Chapter 7). Using digestion by restriction endonuclease isoschizomers, it appears that in vertebrates the majority of cytosines in CG doublets throughout the DNA are methylated. The echinoderms, however, present a clear picture of compartmentalization inasmuch as approximately one-third of their DNA is heavily methylated at cytosines (m^+), whereas the remaining two-thirds are not detectably methylated (m^-) (Bird & Taggart, 1980). While the DNA of some chordates also show this echinoderm-pattern of compartmentalization (*e.g.*, tunicates such as *Ciana intestinalis*, sea squirt), the m^- compartment in the subphylum of vertebrates is, as anticipated, extremely small by comparison (only around 2%) (Cooper *et al.*, 1983). Such observations beg the question of whether such patterns are the only ones to be found in living organisms and whether 5-methylcytosine is the only methylated base of significance in eukaryotic DNA.

11.2   Protista

Turning first to the Protista, 5-methylcytosine is also found in the nuclear DNA of the plant-like Euglenaphyta such as *Euglena* (Brawerman & Eisenstadt, 1964; Brawerman *et al.*, 1962; Ray & Hanawalt, 1964), although its chloroplast DNA is devoid of this modified base. On the other hand, if we move to the dinoflagellates of the Pyrrophyta (Rae, 1976), some have nuclear DNA that contains 5-methylcytosine (*e.g.*, *Exuviaella cassubica* of the Prorocentraceae), and some have DNA containing 6-methyladenine instead (*e.g.*, *Peridinium triguertum* of the Peridiniaceae). Other dinogflagellates, such as *Symbodinium microadriaticum* of the Zooxanthellaceae, appear to have neither 5-methylcytosine nor 6-methyladenine. However it should be noted that these two species do contain hydroxymethyluracil in quite large amounts (3.8–16.4 mol %), which may be the result of modification of thymine residues in specific sequences (Rae, 1976).

The base 6-methyladenine appears again, but this time as the only methyl base, in the DNA of protozoa of the phylum Ciliata, *e.g.*, *Paramecium* (Cummings *et al.*, 1974), *Oxytricha* (Rae & Spear, 1978), and *Tetrahymena* (Gorovsky *et al.*, 1973; Hattman *et al.*, 1978; Blackburn *et al.*, 1983). Ciliated protozoa are characterized by nuclear dimorphism. In the single cells that comprise each organism, two different types of nuclei are found; a diploid transcriptionally inert micronucleus, which functions as a germinal nucleus to maintain genetic continuity between generations, and a transcriptionally active polyploid macronucleus. The *somatic* macronucleus is destroyed in the course of sexual cycle and is replaced by a division product of the newly formed zygotic nucleus produced by meiosis and cross-fertilization of parental micronuclei.

The macro- and micronucleus DNA of *Paramecium aurelia* contains about 2.5 mol % methyladenine. On the other hand, the mitochondrial DNA from this organism contains less than 0.1 mol % of methyladenine (Cummings *et al.*, 1974). In *Stylonychia mytilus* the macronuclear DNA contains 0.176 mol % methyladenine, whereas the micronucleus contains less than 0.04 mol % (Ammerman *et al.*, 1981). The macronuclear DNA of *Tetrahymena pyriformis* also contains 6-methyladenine (0.65–0.8% of adenines are methylated), but again the micronucleus contains only a low level of methyladenine (about tenfold less than in the macronuclear DNA) (Gorovsky *et al.*, 1973).

The methylation of adenine residues of the macronuclear DNA of *Tetrahymena thermophila* has recently been examined at the level of particular genes (Blackburn *et al.*, 1983). For example the genes for ribosomal RNA, which are amplified in the macronucleus, are found in free linear molecules (21 kb) consisting of two ribosomal gene copies arranged as a palindrome. During the amplification, single ribosomal genes (linear molecules of 11 kilobases) are synthesized from the micronuclear ribosomal

RNA gene copies. These molecules, which are not methylated, may serve only as precursors of the final 21 kilobase palindromic form, as they are lost as the macronucleus matures; only the methylated palindromic molecules are retained. Moreover, since the methylation pattern imposed on the palindromic form is retained despite subsequent physiological changes, it is believed that the role of 6-methyladenine residues in these genes is not to regulate transcription but to distinguish DNA molecular forms that are to be retained in the mature macronucleus from those to be rearranged or lost by degradation (Blackburn *et al.*, 1983).

We will consider next the Gymnomycota or slime molds, curious fungi-like protista. Some are decidedly animal-like in their life cycle, while others are more plant-like. In *Physarum polycephalum* (of the Myxomycota), the methylated base detected is once more 5-methylcytosine (Whittaker *et al.*, 1981; Cooney *et al.*, 1984). Moreover the pattern of methylation is quite like that detected in echinoderms. A considerable proportion (20%) of the genome can be isolated in an HpaII-resistant methylated (m^+) fraction (Whittaker *et al.*, 1981). On closer inspection, this m^+ compartment appears to be made up of an abundant methylated family of sequences with an estimated repetitive frequency greater than 2100 derived from a repeated element whose length exceeds 5.8 kilobases (Peoples & Hardman, 1983). The methylation pattern of ribosomal RNA genes in *Physarum* has also been explored. Although some is at CmCGG sites, CGCG sites also appear to be highly methylated. Nevertheless, the transcribed regions of the ribosomal RNA genes are deficient in 5-methylcytosine, but the overall pattern does not change during reversible differentiation of the slime mold (Cooney *et al.*, 1984).

11.3 Fungi

Although fungi are eukaryotic organisms and are primarily multicellular or multinucleate, their cellular organization is fundamentally different from that of animals or plants. It is generally supposed that the DNA of fungi is devoid of methyl bases. Two reports indicate that DNA of *Saccharomyces cerevisiae* (phylum Ascomycetes) lacks modified bases (Guseinov *et al.*, 1972; Wyatt, 1951), and Profitt *et al.* (1984) have recently indicated that any 5-methylcytosine would be present at <1 residue per 3100 to 6000 cytosine residues. In addition to reporting the pyrimidine clusters of *Saccharomyces cerevisiae* to be free of 5-methylcytosine and 6-methyladenine, Guseinov *et al.* (1972) also indicated a similar deficiency in the DNA of the imperfect fungi (*Verticillium dahlia* and *Verticillium lateritium*). Using restriction enzyme analyses, Tamane *et al.* (1983) were also unable to detect any 5-methylcytosine in either *Aspergillus niger* or *Aspergillus nidulans*. These workers also obtained the same results using more sensitive methods, *i.e.*, the use of ^{14}C-methyl-labeled methionine followed by TLC analysis of products (with a sensitivity of 0.1 mol % or 1 methylated site per 10 kilobases) or mass spectrometry (Tamane *et al.*, 1983). This absence of detectable 5-methylcytosine in *Aspergillus* is also confirmed by Clutterbuck and Spathas (1984).

Despite these findings, a recent examination of a wider range of fungi (Antequera *et al.*, 1984) found them to be a heterogeneous group with a 5-methylcytosine content ranging from undetectable levels to low, but detectable levels (*e.g.*, 0.2% and 0.5% of total cytosines in *Sporotrichum dimorphosum* and *Phycomyces blakesleeanus*, respectively). In these latter species, the 5-methylcytosine is located mostly at CpG sequences and the methylated sites are clustered in long tracts (10–30 kbp) separated from essentially unmethylated regions. The methylated compartment (m[+]), which comprises a small fraction (1–11%) of the total DNA sequences, contains at least a specific set of repetitive elements, thus prompting a comparison with the situation already mentioned in the slime mold *Physarum*.

11.4   Animals Excluding Deuterostomia

11.4.1   Porifera, Coelenterata, Mollusca

As already mentioned, 5-methylcytosine is abundant in some of the plant-like protozoa, but 6-methyladenine is the only modified base in the ciliated protozoa. Nevertheless, 5-methylcytosine seems to be the modified base of choice in the DNA of lower animals such as from the Porifera (*Suberites domuncula*), the Coelenterata (*Metridium senile*, sea anemone) and the Mollusca (*Helix pomatia* and *Mytilus edulis*, common mussel) (Vanyushin *et al.*, 1970; Bird & Taggart, 1980). In the case of the DNA of the sea anemone and common mussel, the pattern of methylation has also been examined by restriction endonuclease digestion (Bird & Taggart, 1980) and has been found to be compartmentalized in a fashion very similar to that already reported (see Chapter 7) for the echinoderms (*i.e.*, the echinoderm-pattern).

11.4.2   Arthropoda

Despite these unique occurrences of 5-methylcytosine in lower animals, 6-methyladenine appears along with 5-methylcytosine in some insects. On the other hand it is difficult to generalize about DNA methylation in insects or, indeed, Arthropoda as a whole. Several situations appear to prevail: In some species both modified bases occur; in others both are totally lacking or are present in amounts that are difficult to detect. Considerable effort has gone into the analysis of arthropod DNA for methyl bases in many laboratories. However this is hardly surprising in view of the possible role of methylation in gene expression and the great body of knowledge concerning the developmental biology, particularly of insects such as *Drosophila* (order Diptera, suborder Cyclorrhapha).

Approaches using such methods as HpaII/MspI restriction endonuclease digestion (Bird & Taggart, 1980) indicated the virtual absence of methyl cytosine from the DNA of certain higher insects (from Diptera, such as *Drosophila virilis, Musca domestica,* house fly; *Sarcophaga bullata,* flesh fly, and from Hymenoptera such as *Psithyrus* sp (a bumble bee) (Rae & Steele, 1979). The application of techniques such as high resolution mass spectroscopy (Deutsch *et al.,* 1976), gas chromatography-mass spectroscopy (Singer *et al.,* 1979c), and high performance liquid chromatography (Singer *et al.,* 1977), has confirmed the view that the DNA from *Drosophila,* at least, is extremely deficient in methyl bases. The general sensitivity of these methods determined that less than 0.1% of cytosines could be present as 5-methylcytosine. A similar approach resulted in the same negative results when the DNA of a crustacean (*Artemia,* brine shrimp) was examined (Warner & Bagshaw, 1984). Urieli-Shoval *et al.* (1982) subsequently employed a different approach to improve sensitivity. The DNA of *Drosophila melanogaster* was first nicked with DNase I, then incubated with *E. coli* DNA polymerase I and [α-^{32}P]dGTP. Following this, the ^{32}P-labeled DNA was digested to deoxynucleoside 3' monophosphates by micrococcal nuclease and spleen phosphodiesterase, and the products separated by two-dimensional TLC. Such a nearest neighbor analysis will detect all 5-methylcytosine residues in CpG sequences. Despite the improved sensitivity, no 5-methylcytosine could be detected in DNA from *Drosophila* embryos, larvae, or adults ($\leq$0.1% of CpG sequences or $\leq$1 methylcytosine per 10 kb).

In yet a further attempt to resolve this important question, immunological techniques were employed. Using specifically purified antibodies to 5-methylcytosine, Eastman *et al.* (1980) detected its presence in the bands of polytene chromosomes of *Drosophila melanogaster, Drosophila persimilis,* and *Sciara coprophila* (fungus gnat, Diptera, suborder Nematocera). On the other hand, little was detected in mitotic metaphase chromosomes of *Sciara.* Because DNA in the bands of polytene chromosomes is generally transcriptionally inactive and apparently contains 5-methylcytosine, such qualitative results were consistent with the notion that genes in the bands have been inactivated in a process involving DNA methylation. Indeed later observations using the immunological approach showed a general inverse relationship between RNA transcriptional activity and DNA methylation at puff regions, at least in the polytene chromosomes of *Sciara coprophila.* Despite these interesting observations, a proper quantitative appraisal of the actual level of 5-methylcytosine was still necessary. Achwal *et al.* (1984) have recently reported a method of quantitating the low levels of 5-methylcytosine in *Drosophila* DNA. This involves *spotting* DNA samples in nitrocellulose paper, then using a double antibody technique and staining reaction brought about by biotin-avidin and peroxidase. The stain intensity was then quantitated by photoacoustic spectroscopy. Using this approach, Achwal *et al.* (1984) were able to assess the 5-methycytosine content of *Drosophila* DNA as 0.008 mol % or roughly 1 methylcytosine per 12.5 kilobases. Thus it is possible that methylcytosines are associated with each band of *Drosophila* polytene chromosomes. Another important outcome of the immunological approach was the positive detection of not only 5-methylcytosine in *Drosophila* DNA, but also 6-methyladenine and 7-methylguanine (Achwal, 1983).

A somewhat different approach was used by Adams *et al.* (1979b) to assess the level of methylation of mosquito DNA. Cells of *Aedes albopictus* (Diptera, suborder Nematocera) were labeled with [^{14}C]deoxycytidine, or [^{3}H]adenosine, or with L-[^{3}H-methyl]methionine, and the bases analyzed after hydrolysis of the labeled DNA. This permitted the detection of low levels of not only 5-methylcytosine, but also 6-methyladenine (0.03 mol % of each).

Not all insect DNA is so low in methyl bases. Using high performance liquid chromatography, Deobagkar *et al.* (1982) found the DNA of a mealy bug, from the *Planococcus* species (of the order Homoptera), to contain 2.3 mol % 5-methylcytosine (of which 0.61% occurs as mCpG, 0.68% as mCpA, 0.59% as mCpT, and 0.45% as mCpC). The mCpG accounted for 3.3% of all CpG dinucleotides. Besides 5-methylcytosine, 4 mol % of 6-methyladenine was detected, as well as 2 mol % of 7-methylguanine.

In the DNA of other mealy bugs, Scarbrough *et al.* (1984) also detected high levels of 5-methylcytosine. In *Pseudococcus obscurus,* the level was 1.34 mol % for males and 1.25 mol % for females, and in *Pseudococcus calceofaria* the corresponding levels were 0.68 and 0.44 mol %. On the other hand, Scarbrough *et al.* (1984) were unable to detect any 6-methyladenine or 7-methylguanine in the DNA of these species.

In summary, although the occurrence of methyl bases in insect DNA is still a contentious issue, the available data point to a very low level (0.008 mol %) in the DNA of certain diptera (true flies), such as *Drosophila*. However, it also seems that this very low level is not a feature common to all diptera. For example, higher levels are detectable in representatives of the suborder Nematocera, such as *Aedes albopictus* DNA (0.03 mol %) and *Chironomus plumosus* DNA (0.2 mol %) (Antonov *et al.,* 1962). Even higher levels of methyl bases are encountered in insects of other orders. For example as already mentioned, the DNA of the mealy bugs [order Hemiptera (true bugs), suborder Homoptera] has 5-methylcytosine at 2.3 mol %, although the DNA of adult *Locusta migratoria* (order Orthoptera) has only 0.2 mol % (Wyatt, 1951). Another noticeable feature of insect DNA methylation is that the patterns of methylation may be complex because there are methyl bases other than 5-methylcytosine. As already mentioned, in the same DNA there can also be present similar amounts of 6-methyladenine and even 7-methylguanine. Moreover, as pointed out by Deobagkar *et al.* (1982), the 5-methylcytosine that does occur is *not* predominantly located adjacent to guanine residues. It can be found next to adenine, cytosine, thymine, or guanine residues with almost equal frequency. This of course makes proper assessment of methylation levels in insect DNA with restriction endonucleases difficult. In short, the complexities of the insect methylation patterns still remain to be fully examined before their significance can be assessed. As a postscript maybe it should be stressed that the complete absence of RNA from any of the DNA preparations must be ensured, as this could also be a source of methyl bases.

While representatives of the Crustacea (*e.g., Artemia*) have been examined and seem lacking in methyl bases, it may be premature to pronounce them absolutely free of such bases. As mentioned above, considerably sensitive methods are required before results can be deemed certain. On the other hand, a level of less than one residue of 5-methylcytosine per 50 kilobases in *Artemia* DNA certainly indicates an *extreme* scarcity. However, in the context of Arthropod evolution, it should be pointed out that the Uniramia (the phylum that includes the modern insects) and the Crustacea arose independently more than 500 million years ago, probably from ancestral forms that were already diverged from one another (Manton & Anderson, 1979). Thus, in evolutionary terms, *Artemia* is nearly as distant from insects as both are from the deuterostomes.

11.4.3 Evolution of Methylation Patterns

Despite the problems mentioned above with regard to the insect DNA's, a reasonable generalization is that nuclear DNA methylation ranges from the notably low levels in most insects (insect-pattern) through intermediate levels in nonarthropod invertebrates (echinoderm-pattern) to high levels in vertebrates (vertebrate-pattern). Although an exhaustive survey of animal species has not been carried out, among the species examined so far the methylation pattern does not vary continuously between these extremes. Organisms that are closely related in evolution appear to manifest methylation patterns of a similar type. On this basis, Bird and Taggart (1980) related these three generalized methylation patterns to the accepted phylogenetic relationship between animals, shown in Figure 11.1. The available data, of course, did not indicate which of the three patterns is the ancestral one. Bird and Taggart (1980) proposed that the echinoderm-pattern may have preceded both the insect- and vertebrate-patterns.

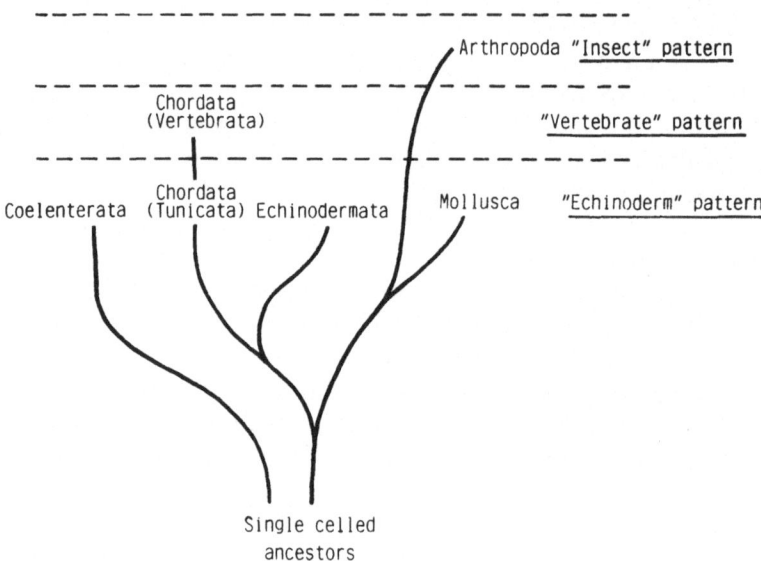

Figure 11.1. Phylogenetic relationships between organisms and the three general methylation patterns.

Whereas the echinoderm-pattern is widely distributed in many phyla (including slime molds and certain angiosperms), the insect- and vertebrate-patterns are only found in single phyla or subphyla (*e.g.*, Arthropoda and Vertebrata, respectively). It could be supposed, therefore, that the echinoderm-pattern was to be found in an ancestor common to these phyla, rather than to assume that the echinoderm-pattern had arisen independently on several occasions. In the case of the vertebrates, there is an additional reason for believing that the present pattern evolved from the echinoderm-pattern. Both echinoderms and tunicate chordates are believed to be closely related to the vertebrate lineage, and both have echinoderm-patterns of methylation. Most crucially, tunicates and vertebrates are members of the same phylum (Chordata).

11.5 Plants

The Chlorophyta (green algae) are of particular interest because they are generally regarded as the group from which the land plants arose. However, whereas the DNA of angiosperms is extensively methylated at cytosine residues only, examination of the nuclear DNA from *Chlamydomonos reinhardi* reveals small amounts of both 5-methylcytosine (0.7 mol %) and 6-methyladenine (0.5 mol %) (Hattman *et al.*, 1978). As already mentioned, the only other organisms so far detected with both modified bases in their DNA are certain bacteria and certain insects. Some of the dinoflagellates of the Pyrrophyta have either 6-methyladenine *or* 5-methylcytosine.

DNA methylation in the plants as a whole has been extensively studied. It is very considerable: Up to 30% of cytosines may be methylated depending on the species (see Shapiro, 1976), and in that sense it is superficially similar to the vertebrate-pattern already discussed. Many of the methylations in plant DNA are found in the dinucleotide sequence CG (Bonen *et al.*, 1980; Deumling, 1981; Gruenbaum *et al.*, 1981a), which of course is the major modified site in animal DNA. It is clear, however, from a study of the nuclear DNA of wheat and of tobacco (Gruenbaum *et al.*, 1981a), that plant DNA contains 5-methylcytosine at CA, CT, and CC sequences as well. While 82% of CG sequences were methylated, 19% of CA sequences, 19% of CT sequences, and 7% of CC sequences were also methylated.

This finding, however, raised a problem of inheritance of plant methylation pattern. The inheritance of the pattern in vertebrate cells strongly suggests the requirement of the symmetrical presence of 5-methylcytosine on both strands of DNA, at least at the sequence CG. Because none of the dinucleotides CA, CT, or CC can accommodate this type of symmetry, Gruenbaum *et al.* (1981a) postulated that methylation at these sites might be based on a trinucleotide symmetry at the sequence C-N-G. Indeed, they provided evidence that 80% of the trinucleotide sequences CTG and CAG were methylated, whereas the nonsymmetrical sequence, CAT, was less than 4% methylated. Despite this critical difference, like the situation in vertebrates, not all the methylatable sites of plant DNA are in fact methylated. Approximately 80% of all CG or $\frac{CAG}{GTC}$ sites, and 50% of $\frac{CCG}{GGC}$ sites are methylated. Thus, as is the case in vertebrates, such incomplete methylation might contribute to gene regulation in plants (Gruenbaum *et al.*, 1981a).

Using restriction enzyme digestion, Naveh-Many and Cedar (1982) were able to show that the sequences of wheat germ DNA that did contain 5-methylcytosine are distributed in clusters and suggest that stretches of unmethylated residues might represent the active positions of the genome. Indeed, in the transcriptional unit for ribosomal RNA from flax, the least methylated region corresponds to the region at which transcription of the ribosomal RNA precursor begins (Ellis *et al.*, 1983).

The origin of the 5-methylcytosine in wheat embryo has been studied at the enzymic level (Theiss & Follmann, 1980). Both L-methionine and L-serine can serve as methyl group donors *in vivo*, and DNA methylation reaches its highest value after the peak of DNA replication at 18 hours after germination. Isolated nuclei from the embryos contain an S-adenosylmethionine-dependent DNA methyltransferase at the stage of germination, although a significant increase is observed at the 22nd hour of germination. Like the vertebrate DNA, methyltransferases the wheat embryo enzyme can be inhibited by analogues of S-adenosylmethionine. DNA methyltransferase is also detected in tissues of dicotyledons, such as pea. From the seedling tissue, a DNA methyltransferase has been extracted (Kalousek & Morris, 1969a), and again S-adenosylmethionine is the donor of methyl groups. The pea seedling enzyme has a sharp temperature optimum at around 30°C. Little is yet known about the sequence specificity of either the wheat embryo or pea seedling enzymes. Bearing in mind the type of methylated sequences detected *in vivo,* it seems essential to establish whether or not plants contain two types of DNA methyltransferase, one recognizing CG sequences and another recognizing CNG-type sequences.

With regard to the kinetics of 5-methylcytosine formation in plants, it has already been indicated that methylation of DNA in germinating wheat embryos follows DNA replication (Theiss & Follmann, 1980). In *Lilium,* methylation occurs during each of the three intervals of DNA synthesis associated with the meiotic cycle (Hotta & Hecht, 1971). Methylation, in fact, can occur at any stage of meiosis except during those stages at which the chromosomes are maximally contracted. Under normal conditions, methylation of cytosines does not occur until after the completion of nascent DNA chains. However, reduction in the level of methylation (with inhibitors such as ethionine or norleucine) does not interfere significantly with either chromosome pairing or chiasma formation during meiosis. Reduction in the level of plant DNA methylation can also be brought about by plant pathogens such as fungi. Guseinov *et al.* (1975) report that the level of 5-methylcytosine in cotton plants (normally 40 mol %) is significantly less when these plants are infected by wilt *Verticillium dahlia* fungus (reduced to 2.3 mol %). Whether this correlates with any changes in nuclear transcription patterns in the cotton plant has not been examined.

Because of the possible involvement of DNA methylation in plant gene control, the methylation of the Ti-plasmid of the pathogen *Agrobacterium tumifaciens* has recently been examined (Gelvin *et al.,* 1983). As has been established in a number of laboratories, a region of this plasmid, called T-DNA, is inserted into the nuclear DNA of dicotyledonous plants during the establishment of crown gall tumors. The T-DNA is transcribed both in *Agrobacterium* and in the plant tumors. Some of these tumors contain a single copy of T-DNA, whereas others contain multiple inserts. Using restriction enzyme analyses, Gelvin *et al.* (1983) were able to demonstrate a lack of methylation of the T-DNA region in the bacterium. In the tumors investigated, at least one copy of the T-DNA is unmethylated. In some tumors containing multiple T-DNA insertions, T-DNA methylation is observed. However, it has not yet been established whether the unmethylated T-DNA inserts are transcribed and the methylated ones not.

11.6 Methylation of Organelle DNA

11.6.1 Mitochondrial DNA

Early investigations (Nass, 1973) indicated the level of methylated cytosine residues in mitochondrial DNA of cultured hamster and mouse cells to be quite low (0.2–0.6% of cytosine residues). Higher values were reported in beef heart mitochondrial DNA by Vanyushin and Kirnos (1974). Positive detection of modified cytosines in mitochondrial DNA has, however, been criticized on the grounds of contamination of the mitochondrial preparations with nuclear DNA. Indeed, Dawid (1974) reported that if ^{32}P-labeled mitochondrial DNA from HeLa cells or from *Xenopus laevis* ovaries was digested with nucleases followed by phosphomonoesterase, no 5-methylcytosine could be detected. The limits of this approach were around 0.23% of cytosines in the case of HeLa mitochondrial DNA and 0.48% of cytosines in the case of *Xenopus* mitochondrial DNA. Using direct digestion with restriction enzymes, Groot and Kroon (1979) were unable to detect 5-methylcytosines in CCGG sequences in mitochondrial DNA from such sources as yeast, *Neurospora,* calf, and rat liver. A similar result was obtained when mitochondrial DNA from wheat was examined using this approach (Bonen *et al.,* 1980).

Despite these findings, the question of mitochondrial methylation remains open. Reis and Goldstein (1983), using human diploid fibroblasts and specific mitochondrial DNA sequence probes (to evade the problem of nuclear DNA contamination), suggest that there may yet be a low level of nonrandomly distributed 5-methylcytosines in mitochondrial DNA. Interestingly, the levels they detect are not very dissimilar to those reported by Nass (1973), but the level appears to decrease with the age of the culture in cells with a finite lifespan (Reis & Goldstein, 1983). Such metabolic variation may preclude definitive answers regarding mitochondrial DNA methylation, at least for the present. Indeed such variations may have profound significance (see below). Pollack *et al.* (1984) also find a similar level of methylation in mouse mitochondrial DNA. From 3 to 5% of CG dinucleotides are methylated, which represents 17 to 29 methylcytosines per mitochondrial DNA molecule. Pollack *et al.* (1984) provide evidence to show that these methyl groups are not distributed at random on the 287 CG dinucleotides of mitochondrial DNA, but neither are particular sequences always methylated.

11.6.2   Chloroplast DNA

Although chloroplast DNA from plants seems devoid of methyl bases
(Baxter & Kirk, 1969; Whitfield & Spencer, 1968; Hattman *et al.*, 1978),
the situation has recently been carefully examined in the green algae
Chlamydomonas. In the vegetative growth stage of this organism the
chloroplast DNA certainly is devoid of methyl groups. Nevertheless,
5-methylcytosine is detectable in the chloroplast DNA of the "female"
gametes of the organism as well as in the zygotes resulting from the fusion
of "male" and "female" gametes. However, 5-methylcytosine was not
detectable in the chloroplast DNA of "male" gametes (Royer & Sager,
1979).

11.6.3   DNA Methylation and Maternal Inheritance

The above-mentioned observations have induced speculation as to the
mechanisms that control maternal inheritance of chloroplast as well as
mitochondrial genes. Maternal inheritance of chloroplast and mitochondrial genes is a fundamental property of the genetics of eukaryotic organisms. It distinguishes chloroplast and mitochondrial genes from those of
nuclear origin. It has been assumed for some time that such a phenomenon resulted from the virtual absence of cytoplasm in the male gamete at
the time of fertilization, *i.e.*, paternal chloroplast or mitochondrial genes
were essentially excluded at fertilization. In the unicellular green algae
Chlamydomonas, maternal inheritance of chloroplast genes has also been
observed (Sager, 1954). However, in this organism both parents contribute their total cell contents to the zygote, thus permitting possible alterna-
tives to the exclusion hypothesis. In the sexual cycle of *Chlamydomonas*,
pairs of morphologically identical haploid cells of opposite mating types
(mt$^+$ and mt$^-$) fuse to form the diploid zygote, which later undergoes
meiosis and gives rise to four haploid zoospores, two mt$^+$ and two mt$^-$.
While the mating type is regulated by a nuclear gene or gene cluster, only
chloroplast genes from the mt$^+$ parent are transmitted via the zygote to all
of the four zoospores. The chloroplast genes from the mt$^-$ parent are not
transmitted. Since this type of transmission pattern is formally identical
with the maternal inheritance of chloroplast genes in higher plants, the
mt$^+$ mating type is considered the *female* and the mt$^-$ type the *male*.

Over the past ten years or so, Sager and her colleagues (Sager *et al.*,
1984) have presented several lines of evidence that support a methylation-
restriction model, similar to that in bacteria, as the molecular basis of
maternal inheritance in *Chlamydomonas*. The model is presented in Figure 11.2. From cytological data of Kuroiwa *et al.* (1982), it appears that
the chloroplast DNA of one parent (mt$^-$) is *degraded,* while that of the

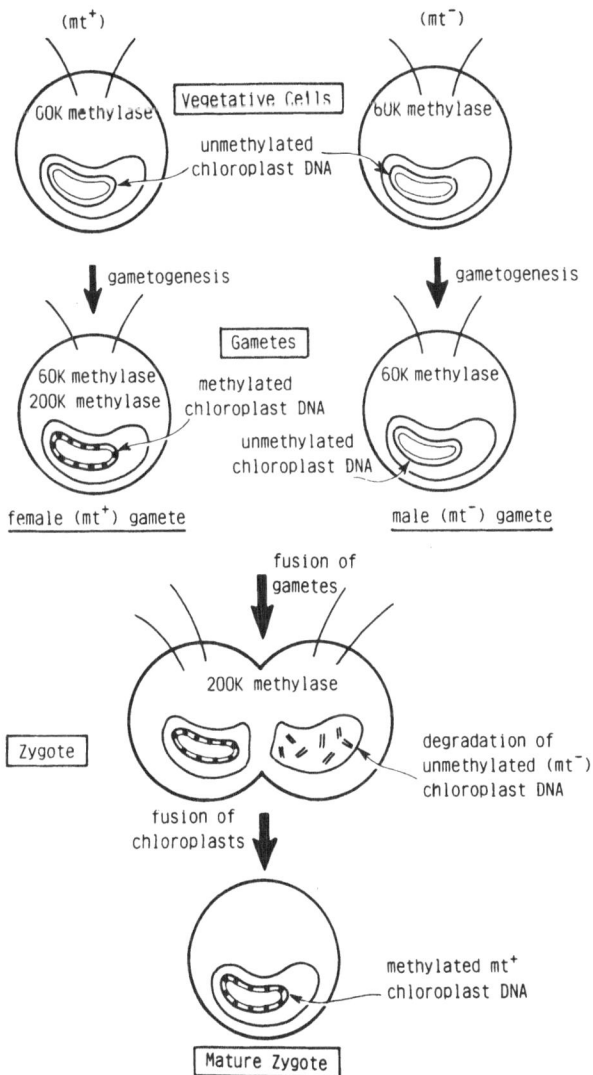

Figure 11.2. A possible model for the maternal inheritance of chloroplast DNA in *Chlamydomonas*.

other (mt⁺) remains intact following zygote formation but *before* chloroplast fusion occurs.

Gametogenesis in *Chlamydomonas* can be induced by suspending the organism in media free of available nitrogen sources. With restriction enzymes, Royer and Sager (1979) looked for indications of methylation of gametic chloroplast DNA's. Apparently, the chloroplast DNA from female gametes (mt⁺) progressively becomes very resistant to HpaII digestion but not to MspI, indicating extensive methylation of CCGG sequences. At 24 hours, 22% of cytosines are methylated. On the other hand, the chloroplast DNA from the male gametes, like chloroplast DNA from both "male" and "female" vegetative cells, remains uniformly sensitive to HpaII.

This approach was then used to evaluate the methylation pattern in a mutant *mat-1* discovered in the mt⁻ population. The *mat-1* mutation carried by males converts maternal inheritance to biparental inheritance (Sager & Ramanis, 1974). In *mat-1* mt⁻ gametes there is evidence of methylation of chloroplast DNA, but this occurs at a very slow rate.

Another mutant *me-1* has been reported (Bolen *et al.*, 1982) in which the chloroplast DNA of vegetative cells (mt⁺ and mt⁻) is extensively methylated (up to 37% of the cytosines). Importantly, this extensive methylation encountered in *me-1* cells appears to have no effect on the phenotype of the cells or on maternal inheritance! Despite this, Sager and Grabowy (1983) subsequently found that during gametogenesis of *me-1* cells, *additional* methylation of cytosines occurs in the *me-1* mt⁺ (female) but not in the *me-1* mt⁻ (male). Thus gamete-specific mating-type-specific methylation still occurs in the *me-1* mutant as in the wild type. The baseline of methylation is simply higher.

To understand more of the mt⁻ gametic methylation, a search for DNA methyltransferases in *Chlamydomonas* was initiated (Sager *et al.*, 1984). Surprisingly, even in vegetative cells there is a methylase (approx. molec. wt. 60,000) that catalyzes the transfer of methyl groups from S-adenosylmethionine to cytosine residues in double-stranded DNA (Sano & Sager, 1980). When gametes and zygotes are examined, however (Sano *et al.*, 1981), there are two forms of methylase evident. Besides a form of molecular weight 60,000 ("60K form"), there is another of molecular weight 200,000 ("200K form"). The 60K form is present in vegetative cells and in gametes of both mating types, but in mt⁺ gametes the 200K form is also present. In zygotes, only the 200K form is apparent. (In *mat-1* mt⁻ gametes, both the 60K and 200K forms can be detected, although the rate of appearance of the 200K form is slow.) The 200K form has been postulated to be a multimer of the 60K form (Sano *et al.*, 1981) and is present only in cells in which methylation actually takes place. *In vitro* the 200K form of the enzyme methylates randomly at CA, CC, CT, and CC, but it methylates both hemimethylated and nonmethylated sites, thus acting as both a maintenance and a *de novo* methylase.

Another question is whether methylation in *me-1* mutant vegetative cells is carried out by the same enzyme that is activated in the gametogenesis of wild-type females and of *mat-1* mutant males, or whether more than one enzyme is involved. A possibility is that the *in vivo* activity of the 60K form may be regulated at the *me-1* locus. Thus the *me-1* mutation could be regulatory, removing an inhibitor or modifying the enzyme. However, in gametogenesis, assembly of the 60K subunit into the 200K form might be expected to be regulated by the *mat-1* gene at the mating-type locus (Sager *et al.*, 1984).

After fusion of mt$^+$ and mt$^-$ gametes to form a zygote, Sager and Lane (1972) were able to show that whereas the chloroplast DNA from the mt$^-$ gamete (identified by prelabeling) soon disappeared, the chloroplast DNA of mt$^+$ (female) origin remained and was further methylated (Burton *et al.*, 1979). This raised the question of a possible *restriction*-type enzyme that would preferentially destroy the unmethylated mt$^-$ chloroplast DNA but leave the mt$^+$ DNA intact by virtue of the specific pattern of methylation imposed upon it by the 200K form of the *Chlamydomonas* methylase (Sager & Ramanis, 1973). [It is suggested (Sager *et al.*, 1984) that methylation in the vegetative cells possibly by the 60K form in *me-1* mutants would not impose such protective specificity.]

Despite the attractiveness of the modification-restriction model, the demonstrable loss of the mt$^-$ chloroplast DNA and the apparent gametic specificity of the chloroplast DNA methylation, some problems remain to be solved. Other groups have found, for example, that while hypomethylation of chloroplast DNA can be brought about during gametogenesis by use of inhibitors such as L-ethionine or 5-azacytidine, this does not correlate with the frequency of maternal transmission of chloroplast genes (Feng & Chiang, 1984). In addition, it must be conceded that no restriction-type nuclease with the required specificity has yet been detected in *Chlamydomonas*. On the other hand, a unique deoxyribonuclease was detected in zygotes and vegetative cells (Burton *et al.*, 1977), but it has not been purified or properly characterized. Nevertheless, crude preparations of the enzyme did cleave adeno 2 virus DNA to yield fragments that were shown to arise from single strand nicks and overlapping gaps made by the enzyme at specific sites. While the activity has some properties similar to a type I restriction nuclease, it nevertheless has not been established whether its action might be blocked by methylation at these sites. Thus the existence of a nuclease in *Chlamydomonas* that would specifically degrade unmethylated mt$^-$ chloroplast DNA and not methylated mt$^+$ chloroplast DNA still remains to be established.

In view of the considerable uncertainty that still remains regarding the possible role of methylation in the maternal inheritance of chloroplast DNA, it is worth examining the mitochondrial DNA situation. To evaluate whether the absence of modification of parental mitochondrial DNA or methylation of oocyte mitochondrial DNA could serve as a basis for maternal inheritance of mitochondrial DNA in mammals, Hecht *et al.* (1984) examined the mitochondrial genome from four meiotic and post-meiotic testicular types and in oocytes of the mouse. Between meiosis and the end of spermatogenesis the number of mitochondrial genomes per haploid genome decreases 8- to 10-fold, with spermatozoa containing approximately one copy of the mitochondrial genome per mitochondrion. As far as could be judged from use of restriction enzymes, there were no gross differences in DNA sequence in the testicular mitochondrial DNA from meiotic cells, early haploid cells, late haploid cells, and spermatozoa. Moreover no methylation differences were detected in mitochondrial DNA from sperm and oocytes following digestion with seven methylation-sensitive restriction enzymes. Thus it would appear that maternal inheritance of the mouse mitochondrial genome is not mediated by a loss or gross alteration of the paternal mitochondrial DNA or by methylation of the oocyte mitochondrial DNA.

12

Has DNA a Role in the Control of Transcription?

12.1 The Dilemma

It is clear that in eukaryotes while the proportion of CpG dinucleotides methylated ranges from extremely small up to 70 to 80%, no genome so far investigated has all its CpG dinucleotides methylated. Could this be meaningful in terms of regulation, or might it merely reflect limiting amounts of DNA methylase that fail to saturate a genome already associated loosely or extremely tightly with other proteins? It is quite possible that the tissue variations reported in Section 8.1 and the undermethylation of transcribing regions (Section 8.3.1.) are simply a consequence of the availability of the DNA for methylation. As such, they tell us little about the function of DNA methylation but more about the different chromatin structures present. The increased levels of methylation found in the mammalian early embryo (Section 8.2) and the ability of cells from such embryos to carry out *de novo* methylation (Section 8.4.3.) may also simply reflect an altered chromatin structure. Alternatively, the early embryo may be a special situation where an altered DNA methylase (Section 5.2.5) is present, which can take advantage of a rapidly changing spectrum of chromosomal proteins to imprint a particular methylation pattern on the DNA. What then could be the function of such an imprinted pattern and how might it be changed?

These are some of the questions that remain to be answered before we will know whether the pattern of DNA methylation can control gene expression or is simply a consequence of such expression.

12.2 The Relationship Between Undermethylation and Expression

It is of course a matter of opinion what weight is given to particular pieces of evidence. It is clear that many genes when they are being expressed show an undermethylation particularly at the 5' end and in the 5' flanking regions. There are other genes that apparently fail to show such a correlation, but this is possibly only a result of the absence of the diagnostic restriction sites in these 5' regions. There are yet other genes that can be undermethylated and yet not expressed, and this has led to the coining of the now hackneyed phrase: *Undermethylation is necessary but not sufficient for gene expression.*

When azacytidine is used to block DNA methylation (Section 8.3) those cells surviving the treatment often show a changed pattern of gene expression. In particular, genes on the previously inactive X-chromosome may be reactivated (Section 8.5.). The changes in expression, however, are not random, despite the fact that more than 80% of the methyl groups may be transiently lost from the DNA. Possibly the lethal effects of most changes mean that only cells with certain limited possibilities will survive. Alternatively, as the treated cells are already "determined" (*i.e.*, in the early embryo they have been imprinted for a certain function in the developed animal), the possible consequences of the drug treatment may already be limited. Thus when β thalassemics are treated with azacytidine (Section 8.3.3), the induction of γ globin gene transcription in erythroid cells may be a consequence not just of the demethylation of the γ globin genes. Expression may require that the chromatin is in an *open* configuration and that *trans*-acting factors specific to erythroid cells are also present. These results do, however, strongly indicate that *methylation of certain genes may disallow their expression.*

It is clear that in at least two cases, changes in DNA methylation occur after changes in gene expression. The chick liver vitellogenin gene is transcribed prior to demethylation following estrogen treatment (Section 8.3.4a), and retrovirus proviral DNA introduced into mouse teratocarcinoma cells is not transcribed, yet methylation does not occur for several days (Section 8.4.3a). In these cases methylation occurs after the changes in chromatin structure associated with changes in gene expression.

To try to resolve the dilemma, an obvious approach is to compare the expression of methylated and unmethylated genes. Several expression systems have been used, but there are difficulties in interpreting the results. Attempts to obtain a satisfactorily methylated gene have met with limited success, and only the experiments using DNA naturally available in a methylated and unmethylated state, *e.g.*, ribosomal DNA (Section 7.3.3.) can be considered reliable. Ribosomal DNA can be isolated in a naturally methylated state from sperm and in the nonmethylated state from oocytes or as a result of cloning, yet the two DNA's are equally well-transcribed in *Xenopus* oocytes. The opposite result is obtained with the *hprt* gene. When isolated from the active X-chromosome, this gene is active when transfected into HPRT⁻ cells, yet the gene isolated from inactive X-chromosomes is inactive. We are assuming (see Chapter 10 for the evidence) that the genes from the active and inactive X-chromosome differ in methylation status. Thus even these *model* results do not lead to a clearcut conclusion. In most other experiments the methylated DNA has been obtained either by using a bacterial modification enzyme, which acts at only a limited number of sites, or by replacing every cytosine by methylcytosine (Figure 8.13). In the former method *significant* cytosines may fail to be methylated (thus apparently showing no effect of methylation), and in both methods cytosines may be methylated that are never methylated *in vivo*. This is important because it has been suggested that particular sequences in the 5′ regions of genes may be protected from methylation *throughout development* (Bird, 1984a; Max, 1984), and that this may be essential for the subsequent expression of the genes. While methylation *in vitro* of such sequences may block subsequent expression, this may not be the mechanism used *in vivo* by the cell. Indeed all potentially active genes may have, bound to their 5′ ends, regulator proteins which (a) allow subsequent expression, and (b) incidentally block access to DNA methylase. These regulator protein-binding sites may or may not contain CpG dinucleotides, but obviously if they do, a base substitution of a methylcytosine for a cytosine could affect the binding of the regulator protein, thereby affecting gene expression.

12.3   Three Classes of Genes

As Bird (1984a) has pointed out there may be three classes of genes in which methylcytosines in control regions are accommodated in different ways. Class 1: The presence of methylcytosine may have no effect on the binding of the control protein, and hence expression can occur whether or not the cytosines are methylated. Class 2: Key sequences may always be maintained in an unmethylated state, perhaps by the permanent presence of a control protein. Class 3: Expression may depend on, or be associated with, the removal of particular methylcytosines from control regions. This may depend on the replication or repair of the DNA in the absence of methylation, but would require the presence of specific proteins to block methylation at specific sequences (or even to direct a demethylase). The presence of these proteins would ensure transcription and demethylation, but whether the latter is necessary or simply incidental is open to question.

These three suggestions all appear to relegate the presence of methylcytosine to the inconsequential. However, the key feature of a maintenance methylase is that it does maintain a particular pattern of methylation once that has been established. The importance of this is that once the presence of specific proteins has established a pattern of methylation, that pattern can be maintained solely by the action of DNA methylase. Thus, if a control protein is present, even transiently, in the early embryo, or at any subsequent time during development, then its particular recognition sites could become demethylated. In any one cell this demethylation may affect only a small number of genes, but these genes would now have the *potential* for expression.

12.4   Ways in Which a Methylation Pattern May Be Established and Changed

There is little firm evidence as to how proteins may affect DNA methylation. It is clear that histones inhibit methylation and that chromatin is a much less efficient substrate for DNA methylase than is naked DNA (Davis *et al.*, unpublished data), and this has led to the suggestion that the rate of methylation may be retarded in nucleosome core regions (Adams & Burdon, 1983). Following replication, if a particular nucleosomal pattern is established before methylation is complete, then certain cytosines may not be methylated and after two rounds of replication certain sites would be completely demethylated, as shown in Figure 12.1. This pattern would depend on a fairly rigid nucleosome phasing established by a control protein bound at the 5' end of the gene. It would explain how certain sites are always methylated, some never methylated, and how some show a level of methylation dependent on the transcriptional activity, *e.g.*, the ovalbumin gene (Section 8.3.3d).

In genes that are very frequently transcribed, the actual transcription apparatus may interfere with methylation throughout the length of the gene, and this may explain the widespread undermethylation of the chick globin genes, the rat albumin gene, and the ribosomal RNA genes in particularly active tissues (Sections 8.3.3c, 8.3.4d, and 7.3.3).

The finding in active chromatin of DNase I-hypersensitive sites, usually just upstream from the start site of transcription, is taken to imply a region of extreme accessibility characterized by the frequent presence of protein-free DNA. Such regions (which may be present as Z-DNA as a result of the loss of nucleosomes; see Chapter 3) might reasonably be expected to be also highly susceptible to damage *in vivo* and, therefore, to be in a state of constant repair. If repair takes place in the absence of methylation (as may be very likely in nondividing cells where DNA methylase is limiting), then the DNA in the region of hypersensitive sites will become demethylated very quickly, as shown in Figure 12.2. Two rounds of repli-

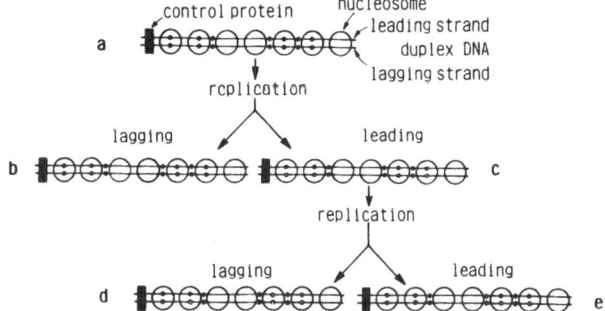

Figure 12.1. The possible effect of nucleosomes on DNA methylation following DNA replication. (•) A methylated site. (○) A potential, but unmethylated, site. (a) Nucleosomes are phased on transcribing DNA by the presence of a control protein. (b) Lagging strand: Following replication the methylation pattern is maintained before new nucleosomes are formed. (c) Leading strand: The daughter strand cannot be methylated where covered by conserved nucleosomes. (d) Lagging strand: As the parental strand lacks methyl groups at some sites, these will not be methylated in the daughter strand. (e) Leading strand: The pattern of (c) will again be produced. Reprinted with permission from Adams and Burdon, *CRC Critical Reviews of Biochemistry 13:*359. Copyright 1982, CRC Press, Inc., Boca Raton, Florida.

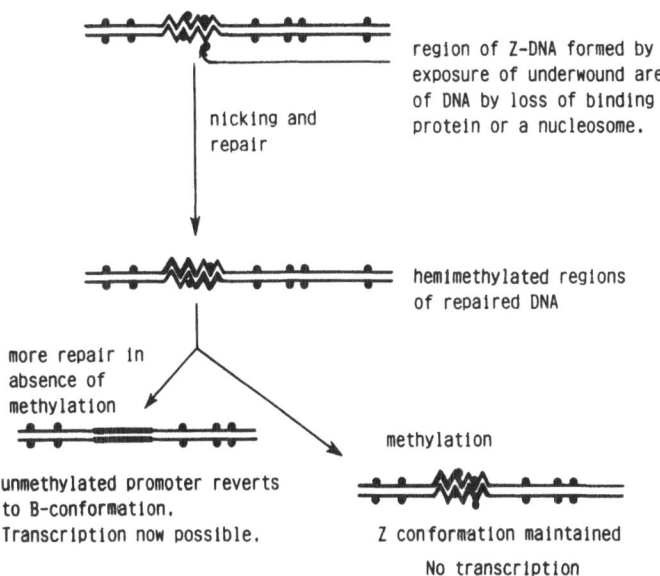

Figure 12.2. Repair of exposed promoter regions (shown here as Z-DNA) may lead to demethylation and a change from the Z to B conformation. This may be the signal that allows transcription to be initiated.

cation will not be required in this case, and the demethylation event is most likely an inconsequential result of the presence of the hypersensitive site. This may be the explanation for the rapid demethylation that occurs to the 5' region of the chick liver vitellogenin gene even in the presence of inhibitors of DNA replication (Section 8.3.4a).

12.5   The Relevance of Z-DNA Formation to the Study of Gene Expression

In Chapter 3 we considered situations that may lead to the formation of Z-DNA and we concluded that although Z-DNA may exist *in vivo,* there is as yet little evidence to involve DNA methylation in its formation. Indeed runs of alternate CpGs are rare and short; although it is possible that a single CpG may exist in a run of alternating pyrimidines and purines, and its methylation may have a controlling influence on whether or not the DNA assumes a Z configuration.

As transcribing genes tend to be undermethylated, it is reasonable to postulate that, as methylation favors Z-DNA formation, a gene activation involving demethylation might be associated with the conversion of a region of Z-DNA to B-DNA. Thus the control region of an inactive gene might contain a short region of a methylated DNA in the Z configuration. Activation may be brought about by a demethylation event destabilizing the Z form of DNA and causing supercoiling in neighboring regions, thereby exposing regulatory sequences, as Figure 12.3 illustrates. However, as the demethylation event, to be specific, would itself require recognition of regulatory sequences, the intervention of these extra steps would only be advantageous if *imprinting* of the signal were required (see above). As suggested by Rich (1982), a gene "promoter" region may exist as Z-DNA and may be stabilized by methylation and by association with Z-DNA-binding proteins rather than with histones. The boundary between B- and Z-DNA may be very sensitive to nuclease action (indeed, it may be the DNase I-hypersensitive site), leading to frequent degradation and repair of this region—a process that must be accompanied by methylation in order to maintain the stability of the Z conformation. Interference with methylation would produce a demethylated region without the intervention of two rounds of DNA replication and cause reversion of the promoter to the B conformation when it could interact with RNA polymerase or other transcription factors (Figure 12.2). Hemimethylated DNA is an intermediate in the repair process, but it is not known whether the hemimethylated promoter would be stable as Z-DNA or would revert to B-DNA. As hemimethylated DNA is very difficult to detect *in vivo,* the change to B-DNA and the accompanying transcription may be apparent before the DNA becomes completely demethylated. That is, before it can be detected by restriction enzyme analysis.

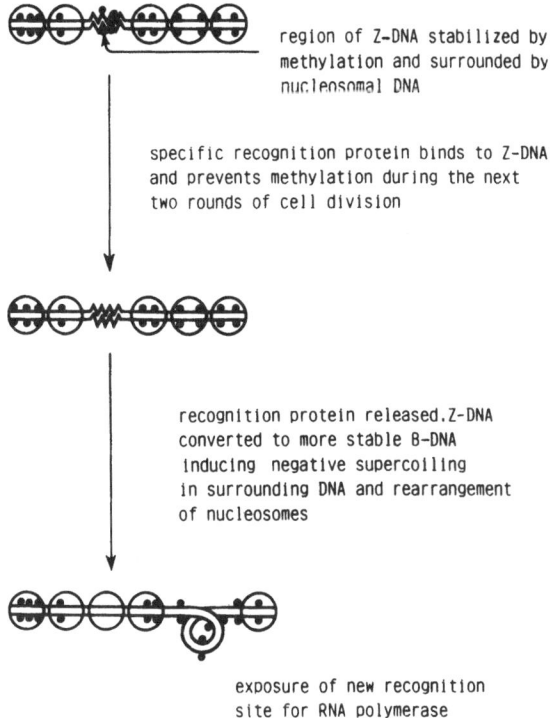

region of Z-DNA stabilized by
methylation and surrounded by
nucleosomal DNA

specific recognition protein binds to Z-DNA
and prevents methylation during the next
two rounds of cell division

recognition protein released. Z-DNA
converted to more stable B-DNA
inducing negative supercoiling
in surrounding DNA and rearrangement
of nucleosomes

exposure of new recognition
site for RNA polymerase

Figure 12.3. An alternative model relating demethylation of DNA to gene activation by means of the induction of supercoiling. (•) Methyl groups in methylcytosine. (〰) Z-DNA.

12.6   A Role in Determination

As we have already mentioned, it would not be unreasonable to suppose that the base change from cytosine to methylcytosine (or the reverse) might affect the binding of control proteins and hence the expression of genes. Whether or not this forms part of a normal process of control over eukaryote gene expression is still very much in doubt. We have concluded in previous chapters, however, that such a control would only make sense if it took advantage of the ability of the DNA methylase to maintain the pattern of methylation in the subsequent absence of the proteins that were initially required to establish the pattern of methylation.

Let us suppose (and there is some evidence for this supposition: McGinnis *et al.*, 1984; Carrasco *et al.*, 1984; Struhl, 1984) that certain groups of genes can be identified by common short sequences of nucleotides near the start site of transcription. These sequences are instrumental in ensuring the coordinated expression of the group of genes. The presence for two cell generations in, say presumptive liver cells, of a specific protein (the determining protein) that recognizes the key sequence and blocks its methylation might ensure the availability for future transcription of those genes whose products are required for specific liver functions, as illustrated in Figure 12.4. In higher eukaryotes at least, the available evidence does not support a role for the majority of the methyl groups in such a determining function. Rather, we believe that most changes in methylation pattern occur as a *consequence* of changes in gene expression. Only a very few of the many methylcytosine groups present in the genome of higher eukaryotes might be important for determination, and the very low levels of methylcytosine present in insects and fungi could well be sufficient to play such a part in determination in these phyla (see Chapter 11). It is, of course, possible that rather than methylation, it

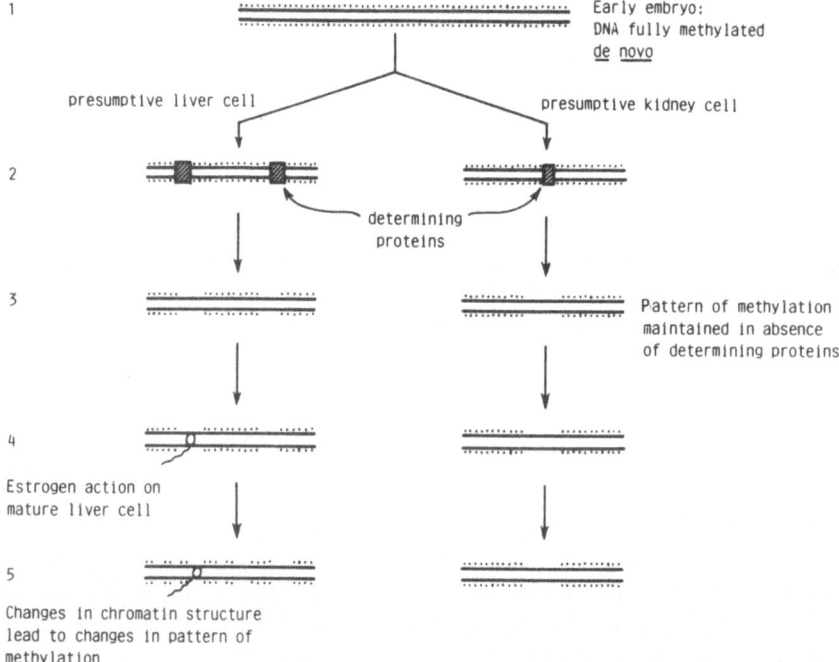

Figure 12.4. Determination and methylation. In the early embryo (1) DNA is fully methylated by *de novo* methylation. Determining proteins then lead to demethylation of certain regions in a tissue-specific manner by replication in the absence of methylation (2). The pattern of methylation is maintained in the absence of the determining proteins (3). Transcription may be induced only in target tissues (4) leading to further changes in the pattern of methylation (5) described in Figure 12.1. The hatched box indicates the determining protein and the circle with a tail represents a transcription complex with its associated transcript. The dots represent methyl groups.

is the chromatin structure itself that plays this role in determination (Brown, 1984). Nevertheless, methylation, which requires just one protein (the methylase) to maintain the programmed state of all genes, has obvious advantages.

Accepting for the moment that sites of methylation are involved in determination, these would not be expected to change their state of methylation in parallel with changes in gene transcription in a given tissue. Thus attention should perhaps be paid to genes that are *potentially* active in a differentiated tissue. This approach might be more fruitful than seeking further inverse correlations between active genes and DNA methylation.

A more complicated mechanism has been pointed out by Konieczny and Emerson (1984). They propose that one to three sites in a *regulatory* locus may become hypomethylated thereby bringing about determination. Expression of differentiated functions would then await the appropriate signal (*e.g.*, cell crowding), which would activate only those genes under the control of the hypomethylated regulatory locus. The level of methylation of the structural genes may alter as a consequence of expression, but it cannot be ruled out that demethylation of a particular site in a structural gene is an additional prerequisite for determination.

Prior to establishment of the pattern of methylation, it would be important for developmental purposes that any previously existing patterns are removed in order to present a *clean slate*. This could occur in the absence of the determining proteins by *de novo* methylation in the germ line or early embryo, such that most CpGs are modified. The determining proteins would then interact with a fully methylated genome and a tissue-specific pattern of methylation would arise slowly as a result of replication and the blockage of methylation at specific sites.

Recent results of Groudine and Conkin (1985) support the idea that early in chick spermatogenesis a wave of *de novo* methylation methylates all the DNA except for certain point sites in active genes. These point sites coincide with nuclease hypersensitive sites in the chromatin. However, in at least one case (the thymidine kinase gene), nuclease sensitivity is lost as the gene is inactivated in mature sperm but is regained following fertilization. It is postulated that the undermethylated site earmarks that paternal gene for early expression in the egg.

Clearly any failure in the activity or specificity of the maintenance methylase could have disastrous consequences. No change in the overall level of cellular DNA methylation may be apparent, but should a vital cytosine here or there gain or lose a methyl group, the *potential* of that cell may change. This may not result in any immediate change in gene expression, but the possibility exists that the cell may later produce, for example, a growth factor that was previously absent and so possibly change itself or some target tissue into a tumor.

One major argument against these suggestions is that in many cell lines the pattern of methylation is far from stable, and viral DNA when it becomes integrated into the host chromosome is frequently methylated *de novo* (see Chapter 8). Whether this represents a breakdown of the normal control system or whether it represents an acceptable latitude in the specificity of methylation is unknown.

Although it is proposed that only a few methylcytosines may be important in the context of determination, the *de novo* methylation in the embryo leads to the methylation of all CpG dinucleotides (except those already blocked by determining proteins or transcriptional control proteins). Those methylcytosine not involved in determination may have other functions (see Chapter 13) but it is, of course, possible that many have no function at all. Indeed, the observation that many methyl CG dinucleotides have been lost by deamination to TpG and CpA without apparent effect supports this possibility (see Section 7.2.).

13

Other Possible Functions of Eukaryotic DNA Methylation

There have been many suggestions concerning the function of DNA methylation in eukaryotes. There is, however, little or no evidence in most cases on which to draw conclusions.

We have reviewed the evidence in the last few chapters for a function of DNA methylation in the control of transcription, and despite a great deal of evidence in this case we have still failed to come to a watertight conclusion. All we have been able to say is that, on balance, the evidence does not favor a role in the control of transcription for the bulk of the DNA methylcytosine.

13.1 Restriction/Modification

Although a nuclease has been isolated from monkey testis that will cleave satellite DNA's at a unique site, there is no evidence to suggest its action is affected by DNA methylation (Brown *et al.*, 1978). Indeed, micrococcal nuclease cleaves certain sequences in specific DNA molecules very much more frequently than other sequences (Dingwall *et al.*, 1981; Nelson *et al.*, 1979), but this is not considered to be a restriction endonuclease. Similarly, yeast cells have site-specific endonucleses (Watabe *et al.*, 1983, 1984), but there is no evidence that their action is affected by DNA methylation or that they play a part in a growth restriction phenomenon.

13.2 Mismatch Repair

The second function for which there is strong favorable evidence in prokaryotes (Section 6.2.2) may also be important in eukaryotes. There is a
similar, even more extensive, lag between DNA synthesis and methylation in eukaryotes, but as yet no evidence is forthcoming that a mismatch
repair system based on delayed methylation operates in eukaryotes.

13.3 Recombination

There is considerable speculation at present that methylation of DNA
may have a function in the control of recombination (see Section 7.4).
Satellite and foldback DNA sequences, which are sites of recombination
and amplification, are highly methylated, and one suggestion is that the
methylcytosine might inhibit the action of a nuclease required for recombination. During meiosis when satellite DNA is undermethylated (Section
2.3.2), the frequency of recombination is markedly enhanced. In cases
where specific recombination events are known to occur (*e.g.*, in the
immunoglobulin gene region), a number of methylated CpG dinucleotides
are present near the site of DNA rearrangement, but no regular pattern
has emerged as yet.

An unusual DNA modification (but not methylcytosine) exists between
the variable surface glycoprotein(VSG) genes of trypanosomes and the
telomeric sequences (Raibaud *et al.*, 1983; Pays *et al.*, 1984; Bernards *et
al.*, 1984). The modification is generally associated with the inactive
genes, but only those in telomeric positions. Whether the modification is
related to transcriptional activity or to the transposition events involved
in activation is unknown. Also unknown is the exact nature of the modification and its possible relationship to more common methylation events.
It is possible that, in an analogous way to the phage mu modification
(Section 6.3), the modified cytosine in the telomeric VSG genes is a hy-
permodified methylcytosine.

13.4 Biological Clocks

Holliday and Pugh (1975) proposed a mechanism, shown in Figure 13.1,
whereby progressive methylation of a repetitive sequence would act as a
biological block counting cell divisions. The mechanism required the initial action of a *de novo* methylase (methylase I) to add one methyl group
followed by the action of a methylase (methylase II) that adds two methyl
groups to the repeat adjacent to the hemimethylated site and also one to
the hemimethylated site (*cf.* maintenance methylation). At replication the

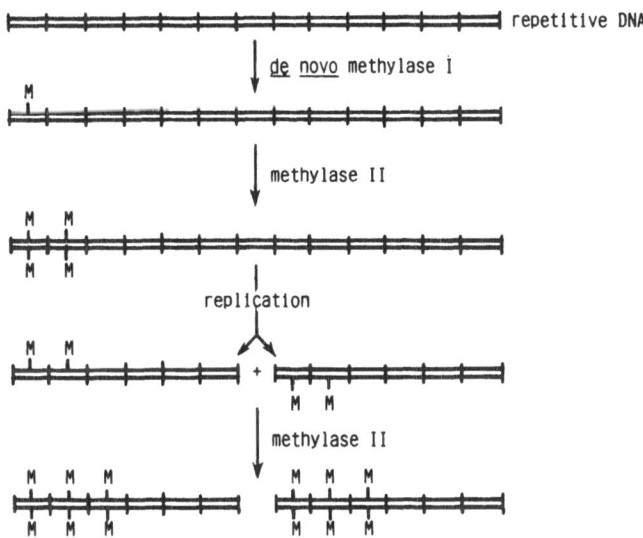

Figure 13.1. A biological clock. Spread of methyl groups along a repeating sequence. Once initiated by methylase I (which adds only one methyl group), methylation is spread by methylase II acting on the repeat adjacent to the hemimethylated site. (From Holliday & Pugh, 1975. Copyright 1975 by the AAAS.)

repeat is transformed into hemimethylated DNA recognized as a substrate for methylase II, which again adds two methyl groups to the adjacent repeat and so on. This leads to the spread of methylation from one repeat to the next at each round of DNA replication. If different numbers of repeating sequences precede different genes, then the genes could be activated (or deactivated) at different times in development as the methylation spreads to them. This mechanism could also explain the spread of the region of inactivation along an X-chromosome (see Chapter 9) and the switch from synthesis of fetal to adult globin (Wood *et al.*, 1985).

Scarano's hypothesis involving the sequential methylation of cytosine followed by the subsequent deamination of the methylcytosine to thymine (1969) is a somewhat similar mechanism by which it was envisaged that genes may be switched on and off during development (see Section 7.2). There is no evidence that either of these mechanisms of measuring time operate in eukaryotes (or prokaryotes), yet the undermethylation of satellite DNA in sperm and eggs (Section 7.3.2) is not inconsistent with the ideas of Holliday and Pugh (1975).

13.5 Initiation of DNA Replication

The region of the origins of DNA replication may be enriched in CpG dinucleotides (as they are in SV40 DNA), and in chromosomal DNA these regions may normally have a high concentration of methylcytosine (*cf.* the *E. coli* origin, Section 6.2.3). The origin DNA, which in G1 phase is methylated in both strands, is the site of attachment of the complex that initiates DNA replication at the beginning of S phase. The continued presence of some of these initiating proteins or the binding of proteins that specifically recognize hemimethylated DNA may block the rapid methylation of origin DNA. The resulting hemimethylated DNA may no longer be able to serve as a site of initiation of replication. In this way reinitiation events are blocked, and only one round of replication occurs at each cell cycle. It is suggested that the blocking proteins are removed at the end of S phase, and the origin is methylated. As the initiation proteins are no

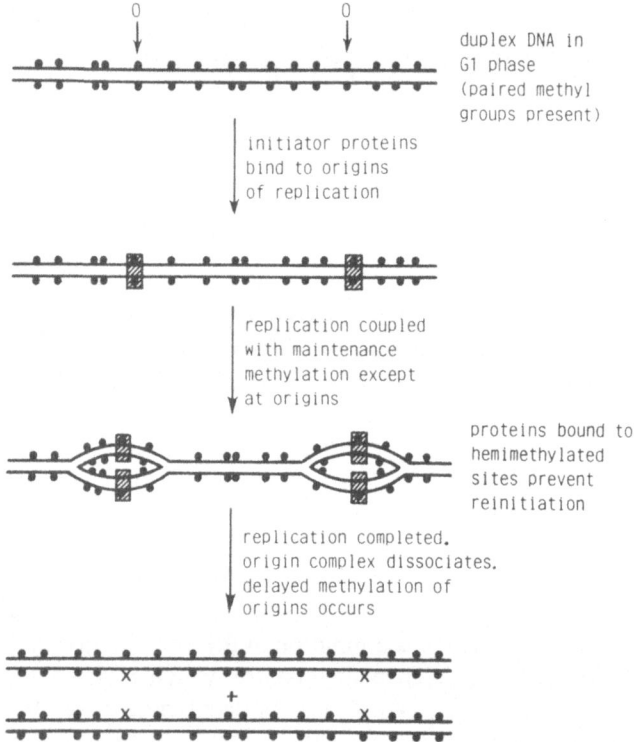

Figure 13.2. Regulation of initiation of DNA replication. Initiation proteins present at the origin of replication may block methylation. The continued presence of these or other proteins firmly bound to the hemimethylated origin prevents reinitiation of replication. The blocking proteins are only released after replication is over when initiation proteins are no longer present. Premature release (as may happen in the continued presence of inhibitors of DNA synthesis, such as hydroxyurea) may lead to reinitiation. Methyl groups are indicated by closed circles and origins of replication with an 0. X marks the site of delayed methylation previously covered with initiation proteins (hatched boxes).

longer present, reinitiation does not occur, as Figure 13.2 illustrates. This model proposed by Taylor (1978) seeks to explain the observation that a significant proportion of DNA methylation in eukaryotes takes place up to several hours after DNA synthesis (Section 2.2b). Otherwise there is no supporting evidence and Woodcock *et al.* (1984) have shown that all renaturation classes are subject to delayed methylation.

13.6 Chromosome Inactivation

Sager and Kitchin (1975) proposed that DNA methylation may help to explain the mechanism of selective inactivation and/or elimination of chromosomes or parts of chromosomes. As an example they suggest that modification enzymes that are present in the female germ cells but not in the male germ cells could *imprint* one set of chromosomes for *subsequent* condensation or elimination in the fertilized egg. Sager and her group have produced considerable data to show that this mechanism acts to ensure maternal inheritance of chloroplast DNA in *Chlamydomonas* (see Section 11.7.1). Others (*e.g.*, Feng & Chiang, 1984) do not agree with Sager's interpretations, and selective oocyte methylation cannot explain the maternal inheritance of the mouse mitochondrial genome (Hecht *et al.*, 1984). Nonetheless, Sager and Kitchin (1975) describe a whole series of phenomena in plants, insects, and unicellular eukaryotes that could be explained by methylation being the first step in a process leading to chromosome inactivation or elimination. In mammals they suggest that X-chromosome inactivation may be similarly explained, and they suggest that the preferential chromosome elimination found in interspecies cell hybrids may result from differential methylation of the two sets of chromosomes.

An observation that supports this hypothesis is the finding that satellite DNA in mammals becomes heavily methylated in early development at a time when its proposed function in meiotic recombination is over. This loss of function is associated with heterochromatinization similar to that which typically precedes chromosome loss in other systems. However, it cannot yet be concluded that the methylation precedes the inactivation process, for it is equally likely to be a passive consequence of the change in chromatin structure.

Although attractive, this model requires that a considerable amount of work be done on the various systems considered, before it can be accepted. Levels of methylation of the different chromosomes in the different cells must be established, and DNA methylase activities must be monitored. It must be shown that the elimination or condensation of chromosomes is dependent on their methylation, and a series of mutants should be obtained that show correlated changes in methylation and inac-

tivation. Finally, it is important to define the stages of chromatin condensation, their dependence on the state of DNA methylation, and the properties of the nuclease(s), which presumably acts to initiate the process of
elimination. Even then, as is clear from the work with *Chlamydomonas,* it
may be difficult to satisfy the scientific community that methylation plays
a definitive role in chromosome inactivation.

14

DNA Methylation in Perspective—A Summing Up

In this book we have presented basic information about DNA methylation. Much of this, especially with respect to eukaryotes, is intricate and complicated and unfortunately does not allow firm conclusions to be drawn as to the function of DNA methylation. Indeed, it could be that there may not simply be *one* function for DNA methylation, but several. On the other hand, it may not be necessary to propose that all methylated bases have a function. After all there are quite long stretches of apparently redundant DNA in eukaryotic cells, and it may very well be that a DNA methylase is actually required to act only on a limited number of bases, but by chance also methylates several others in a similar environment. In bacteria, restriction-modification involves specific base methylation, yet these account for only a minority of the methylated bases actually present in the bacterial DNA. In higher eukaryotes there is as yet no evidence for any restriction-modification system involving base methylation. Indeed it is arguable that methylation is unimportant, because some insects and fungi have extremely low levels of methylated bases. This of course overlooks the possibility that it is just this low level that is required to fulfill some vital role. In this connection it is notable that substantial methylation occurs in *Neurospora crassa* when transforming DNA becomes stably amplified (Bull & Wootton, 1984). It will be important in this system to establish how the *N. crassa* DNA methylase discriminates between the different fractions of DNA involved. This of course is also a key problem in organisms that show distinct compartments of methylated and unmethylated DNA.

While considering functions for DNA methylation, one must not forget the extremely high levels of cytosine methylation found in higher plants.

While the effects of such high levels are as yet unknown, plants may well provide interesting future experimental systems for the study of DNA methylation. It is difficult to imagine such high levels of methylation to be without function. A possible explanation may lie in the fact that plants are notoriously susceptible to polyploidy, whereas some insects that have extremely low levels of methylation have low levels of cellular DNA compared with higher animals and plants. It may be that methylation occurs *primarily* on *redundant* DNA (although clearly methylation may be tolerated in other regions).

The recent surge in interest in DNA methylation has arisen from the realization that in eukaryotes transcribed genes are undermethylated. We have reviewed the evidence for this and conclude that the situation is seldom as clearcut as various authors have led us to believe. One of the clearest examples, however, is the finding that the *hprt* gene from an inactive X-chromosome is poorly expressed when transfected into cells, in comparison to the same DNA from an active X-chromosome. It is difficult to see how this difference could arise other than as a result of DNA methylation or other covalent modification. This, together with other experiments considered in Chapters 7 to 12, provides convincing evidence that methylation *can* affect gene expression. However, a key question that must be asked is whether methylation *in vivo* is ever used to control gene expression. Future experimentation will have to be stringently directed toward answering this question.

When considering the experimental significance of CpG methylation in relation to gene expression, it should be emphasized that many vertebrate genes have short stretches of G+C-rich DNA in their 5'-coding and noncoding regions. Such regions are not deficient in the CpG dinucleotide, in contrast with much of the remainder of the mammalian genome. It has been suggested that such regions are never actually methylated in the cell and are thus not susceptible to mutation as a result of any deamination of 5-methylcytosine to thymine. If this is the case, then experimental methylation of such regions *in vitro* would represent modifications that would not normally occur *in vivo,* thus limiting their biological significance. Such an argument however cannot be applied as an explanation for the low efficiency of transcription following transfection of DNA from the inactive X-chromosome or of inactive retroviral provirus sequences. The point that requires emphasis here is that these gene sequences have been subject to no more than the normal cellular methylation processes. Nonetheless, these genes, once inactivated, are never normally expressed *in vivo*. X-inactivation is an essentially irreversible process that may be accompanied by methylation *either* to facilitate compact chromatin organization, *or* as a consequence of chromatin taking up a compact organization. In a similar manner, DNA methylation of integrated viral DNA may be part of a defensive mechanism of the cell—a mechanism that is largely successful in inactivating expression by leading to the packaging of the

DNA sequence in an inaccessible form. (The methylation could, of course, again be a consequence of such packing.) It is only rarely, when the mechanism fails and some viral genes are expressed, that cell transformation is brought about.

Post synthetic modification of DNA effectively changes the nucleotide sequence and may, thereby, affect DNA protein interactions. As discussed in Section 9.3, certain CGs in the *g6pd* gene are always methylated when the gene is transcribed, while other CGs are methylated when the gene is X-inactivated. This may be the first example in eukaryotes of methylation of different regions of a single gene serving two different functions.

In this book we have also reviewed the properties of different types of DNA methyltransferase on the basis that we can probably learn something about the function of methylation from their mechanisms of action and levels of activity within the cell. With the cloning of bacterial DNA methylases, it has become possible to study the effects of aberrant methylation patterns induced by introducing the bacterial enzymes into eukaryotic cells. The DNA methylase gene from *B. sphaericus* has been introduced into and is expressed in yeast cells (Feher *et al.*, 1983). The enzyme, which methylates the internal cytosine in the sequence GGCC, brings about a nonrandom methylation of yeast DNA. Some yeast genes are heavily methylated, but the overall level of methylation of total yeast-cell DNA is low. Similar results are obtained when the *dam* methylase is expressed in yeast cells (Hoekstra & Malone, 1985). This implies that over most of the genome, the DNA is protected in some way from such methylation. The system thus provides us with the possibility of investigating the factors governing this differential methylation of DNA regions.

Some comment should be made about demethylation. While no convincing evidence has yet been provided, either in prokaryotes or in eukaryotes, for the existence of demethylating enzymes that would remove the methyl group from methylcytosines, demethylation in eukaryotes has apparently been observed to occur in the absence of DNA replication. In relation to the vitellogenin gene this "demethylation" occurs in a DNase I-sensitive region of the chromatin, and we suggest that such demethylation may occur by *repair* of DNA in the absence of methylation. Such a phenomenon may be general and could explain the regular finding of undermethylated DNA near the 5' ends of expressed eukaryotic genes and possibly the changes in patterns of gene expression that are reported to occur in the absence of DNA replication (see Chiu & Blau, 1984).

While considerable recent research effort has been expended in attempts to clarify the links between DNA methylation and gene transcription, future research will have to look at the other potential functions suggested for DNA methylation, such as in the regulation of recombination events.

Despite considerable progress over the last few years, there still is a lot to be learned about the phenomenon of DNA methylation. How have the methylase genes envolved, and what role have the methylases played in shaping the evolution of animal and plant genomes? Can the process of DNA methylation be manipulated to advantage? For example the DNA from tumors, especially metastic tumors, often has lower levels of cytosine methylation than DNA from corresponding normal cells. Are these lower levels a consequence of occasional failures of the methylase to carry out maintenance methylation efficiently? Such a failure may occasionally lead to demethylation of certain critical sites, thereby promoting expression of the genes, say for a particular growth factor. Enhanced production of such a factor may stimulate the uncontrolled growth of that cell or a different target cell or tissue. Although highly speculative, such suggestions indicate that a continued search for drugs or other agents capable of controlling the activity of cellular DNA methylase might yet be highly rewarding.

Appendix: Methods of Estimation of Minor Bases in DNA

A.1 Introduction

Although some of the methods used have already been described in the text, those methods used, in particular, to estimate 5-methylcytosine in DNA have been collected in this Appendix for comparative purposes.

The direct methods involve hydrolysis of the purified DNA to its constituent bases, nucleosides or nucleotides, followed by their separation by some chromatographic method. Those in common use today involve high performance liquid chromatography (HPLC). The amount of individual bases can be quantitated optically, and if the DNA is prelabeled, they can also be quantitated on the basis of their radioactivity.

The amount of methylcytosine in a DNA sample can also be measured by reaction with a specific antibody or by its susceptibility to a particular restriction endonuclease. By contrast, the number of unmodified and potentially methylatable sites can be assayed by measuring the efficiency of a DNA sample as a substrate for a DNA methylase.

The proportion of a particular dinucleotide that is methylated can be estimated by a modified nearest neighbor analysis followed by separation of the 3' mononucleotides.

The Maxam and Gilbert (1980) method of sequencing DNA also indicates when a methylcytosine occurs, but seldom is there any opportunity to apply the method to genomic as opposed to cloned DNA. However, recent advances have allowed the method to be applied to preparations of bulk DNA when suitable probes are available, and this should lead to a new burst of publications on the distribution of this minor base (Church & Gilbert, 1984).

A.2   Isolation and Hydrolysis of DNA

DNA is readily isolated by lysing cells in a solution containing sodium dodecylsulfate. The protein is precipitated using phenol and/or chloroform, and the DNA precipitated with 70% ethanol. Traces of protein and RNA can be removed enzymically or RNA can be selectively hydrolyzed with alkali and the DNA reprecipitated. Purity can be assessed by spectral analysis (Cantoni & Davies, 1966; Maniatis *et al.*, 1982).

Chemical hydrolysis of DNA can be achieved by hot acid treatment. For instance, treatment with 98% formic acid for 90 minutes at 170°C produces complete hydrolysis of the DNA to its constituent bases but is unfortunately associated with the deamination of about 0.4 to 1.0% of the cytosine and methylcytosine and 7.5% of the guanine, as shown in Figure A.1 (Ford *et al.*, 1980; Adams & Clark, personal observation). Eick *et al.* (1983) found no evidence for deamination when the hydrolysis was done for only 30 minutes. Enzymic hydrolysis of DNA produces little deamina-

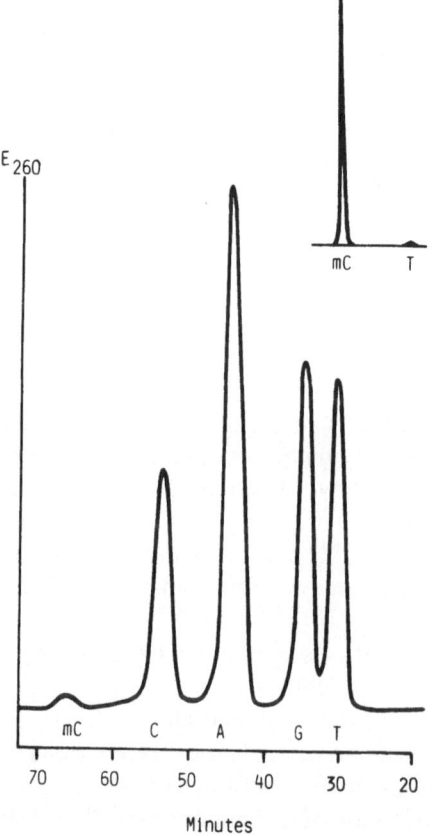

Figure A.1. Base separation on Aminex A6 (Bio-Rad Laboratories). The main trace shows the E_{260} absorbance of the eluate from an Aminex A6 column loaded with a formic acid hydrolysate of salmon testis DNA. The separation takes about 1 hour with the column running at 1 ml per minute at 60°C. The buffer used is 20 *mM* ammonium carbonate pH 10.0. A, adenine; C, cytosine; G, guanine; mC, methylcytosine; and T, thymine. The insert shows the small amount of deamination occurring when a sample of pure methylcytosine is subjected to similar hydrolysis prior to fractionation.

tion (Ford *et al.*, 1980). Hydrolysis with pancreatic DNase I and venom phosphodiesterase yields 5' mononucleotides, while use of micrococcal nuclease and spleen phosphodiesterase yields 3' mononucleotides. Bacterial alkaline phosphatase can be used to remove the phosphate group prior to fractionation of the nucleosides.

A.3.1   Base Analysis

The bases produced by chemical hydrolysis of DNA can be separated by paper or thin layer chromatography, but these methods are not much used these days because they are time consuming and yield inferior separation (Adams & Burdon, 1983). HPLC columns include Partisil SCXK 218 (Reeve Angel), Durrum-DC-1A, and Aminex A6 or A10 resin (Bio-Rad Laboratories). Figure A.1 shows a base separation on Aminex A6 resin. Optical analysis of such separations requires a minimum of about 30 μg DNA, but if the DNA is prelabeled, much smaller amounts can be applied.

DNA can be prelabeled *in vitro* or *in vivo*. In the DNA methylase reaction, radioactivity is transferred from [^{3}H-methyl] S-adenosyl methionine to DNA, and when the mouse DNA methylase is used the only base labeled is 5-methylcytosine.

In vivo DNA can be labeled in the base or its substituent methyl group. Incubation of cells with or injection of animals with [^{3}H-methyl]-methionine leads to transfer of the methyl group via AdoMet to produce methylcytosine. However, unless precautions are taken, the radioactivity enters the 1-carbon pool and labels purines and thymine as well. This incorporation can be limited by the presence of aminopterin (to block *de novo* synthesis of purines and thymidylate), thymidine and hypoxanthine (to supply exogenous precursors), and sodium formate (to flood the 1-carbon pool with nonradioactive folate derivatives). This method of labeling the methylcytosine does not give the proportion of cytosines methylated but only a calculated value of moles methylcytosine per gram DNA.

More convenient ways of prelabeling involve the use of ^{14}C deoxycytidine or 6[^{3}H] uridine. The former is incorporated only into DNA, but labels thymine as well as cytosine and methylcytosine. It is also very expensive and certainly not very suitable for short-term incubations. 6[^{3}H] uridine is incorporated largely into RNA, but a sufficient proportion is incorporated into the DNA pyrimidines, as Figure A.2 illustrates. Following base separation of DNA labeled with either ^{14}C deoxycytidine or 6[^{3}H] uridine, the proportion of cytosines methylated can be calculated from

$$\frac{\text{cpm in methylcytosine} \times 100}{\text{cpm in cytosine} + \text{cpm in methylcytosine}}$$

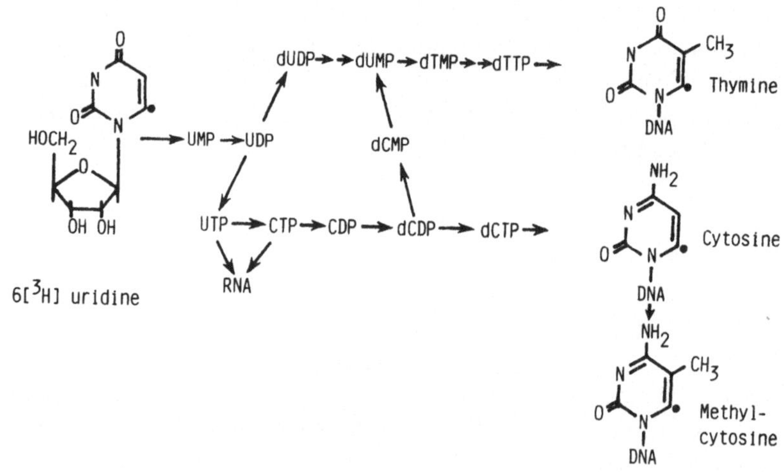

Figure A.2. Incorporation of 6[³H] uridine into DNA. The dot indicates the position of the tritium atom. The pathway shows how radioactivity can be incorporated into RNA and into thymine, cytosine, and methylcytosine in DNA.

Razin and Cedar (1977) have developed a mass spectrometric method of base analysis, which involves hydrolysis of the DNA with trifluoroacetic acid. Here the amount of methylcytosine (m/e 125) is compared with thymine (m/e 126).

A.3.2 Nucleoside and Nucleotide Analysis

Deoxyribonucleosides can be separated on a column of Nucleosil 5C18 (Mackery and Nagel Co.) using 0.1% phosphoric acid (Wagner & Capesius, 1981), or on reversed phase μ Bondapak columns (Waters Assoc., Inc.) using buffers containing first 2.5% methanol followed by 8% methanol (Kuo *et al.*, 1980). About 100 μg DNA are normally used, but results can be obtained with as little as 5 μg. When two columns are used in series, the deoxyribonucleosides are sufficiently well separated from ribonucleosides that moderate levels of contaminating RNA do not interfere. Nucleosides can also be fractionated on Aminex A6 (Bio-Rad Laboratories) using 0.4 *M* ammonium formate pH 4.65 (Thiery *et al.*, 1973).

Recently Eick *et al.* (1983) have reported a good separation of dCMP and methyl dCMP on a μ Bondapak C18 column (Waters Assoc., Inc.) using 5 μM tetrabutylammonium phosphate pH 7.5. However, as described in Chapter 7 and Figure 7.4, these can also be separated by two-dimensional thin layer chromatography. Separation is also achieved by prolonged high voltage electrophoresis (Figure 5.5.)

A.4   Use of Antibodies to Methylcytosine

Antibodies can be raised against methylcytosine and then used to identify the minor base in samples of DNA immobilized on nitrocellulose filters. Sano *et al.* (1980) incubated the immobilized DNA with the rabbit antiserum and then with ^{125}I-labeled goat anti-rabbit IgG. They were able to detect the single methylcytosine base present in as little as 40 ng ϕX174 DNA. Achwal *et al.* (1984) used a biotinylated goat anti-rabbit second antibody and then linked this through avidin to biotinylated peroxidase. The nitrocellulose was then stained for peroxidase and scanned by photoacoustic spectroscopy to give a signal that was directly proportional to the methylcytosine content. Using this method, they report the presence of one methylcytosine residue in about 12,500 bases of *Drosophila* DNA.

A.5   Restriction Enzyme Digestion

Table A.1 lists a number of so called "CG" restriction enzymes and indicates their sensitivity to cytosine methylation as far as is known. Throughout this book, reference has been made to the use of the isoschizomers HpII and MspI to determine whether or not particular CG dinucleotides are methylated (see Chapter 8 and Figure 7.12).

As described in Figure 8.1, when high molecular weight DNA (greater than 50 kbp) is cleaved with MspI it produces on electrophoresis a profile from which the weight average molecular weight can be obtained (Singer

Table A.1. Restriction Enzymes Containing a CG in Their Recognition Sequence-effect of Cytosine Methylation[a]

mCG Sequences			Enzymes blocked by methylation of CG	Enzymes not affected
C	mCG	G	HpaII, HapII	MspI
	mCG	CG	ThaI	
CG	mCG		ThaI	
G	mCG	C	HhaI, HinPI HaeII[b]	
CY	mCG	RG	AvaI	
GR	mCG	YC	AosII, AhaII	
RG	mCG	CY	HaeII, NgoI	
CC	mCG	GG	SmaI	XmaI
CGAT	mCG		PvuI, XorII	
CT	mCG	AG	BsuMI, XhoI	
GT	mCG	AC	SalI	
T	mCG	A		TaqI, TthI

[a] Data from McClelland (1981) and McClelland and Nelson (1985). Y, pyrimidine; R, purine.
[b] HaeII recognizes only a subset of HhaI sites.

et al., 1979a). This agrees reasonably well with the result predicted from base composition and nearest neighbor analysis, assuming a random sequence distribution.

When HpaII is used instead of MspI, a higher value is obtained for the weight average molecular weight from which the proportion of CCGG sequences methylated at the internal cytosine can be calculated.

Besides requiring high molecular weight DNA for the initial digestion, one has to assume that there is a random distribution of fragment sizes in order to apply the relation that the number average molecular weight is half the weight average molecular weight (Gama-Sosa *et al.,* 1983b; Tanford, 1961). As this does not hold for eukaryotic DNA, the conclusions from this method can be only approximate.

An alternative method devised by Cedar *et al.* (1979) involves labeling the 5' ends of MspI cuts in DNA, *i.e.,* the cytosine or methylcytosine in the sequence C(m)CGG is given a radioactive 5' phosphate. On hydrolysis to the 5' phosphates and separation of dCMP and methyl dCMP by two-dimensional thin layer chromatography, an unequivocal measure of the proportion of CCGGs that are methylated can be obtained (see Figure 7.4).

A.6 Nearest Neighbor Analysis

The above procedure can be made more general if nicks are introduced into DNA at random. This is not entirely possible, however, but an approximation can be obtained by nicking with DNase I or by sonication. The nucleotide on the 3' end of the nick can be labeled by incubation with DNA polymerase and a single α^{32}P-labeled 5' triphosphate (see Figure 7.3). On hydrolysis to the 3' monophosphates, this modified nearest neighbor analysis can indicate the proportion of the four CpN dinucleotides that are methylated (see Figures 5.5 and 7.4). Pollack *et al.* (1984) estimate that this method can detect one methylcytosine in 1000 of each of the four dinucleotides CpG, CpA, CpC, or CpT.

A.7 Maxam-Gilbert Sequencing

In dilute buffer, hydrazine interacts with pyrimidines in DNA leading to opening up of the pyrimidine ring, as shown in Figure A.3. Subsequent treatment with piperidine breaks the sugar phosphate backbone on both sides of each opened ring, and the fragments produced can be sized by gel electrophoresis (Maxam & Gilbert, 1980). In high salt conditions, only the cytosine is cleaved. Figure 7.2 shows a gel fractionation and illustrates the

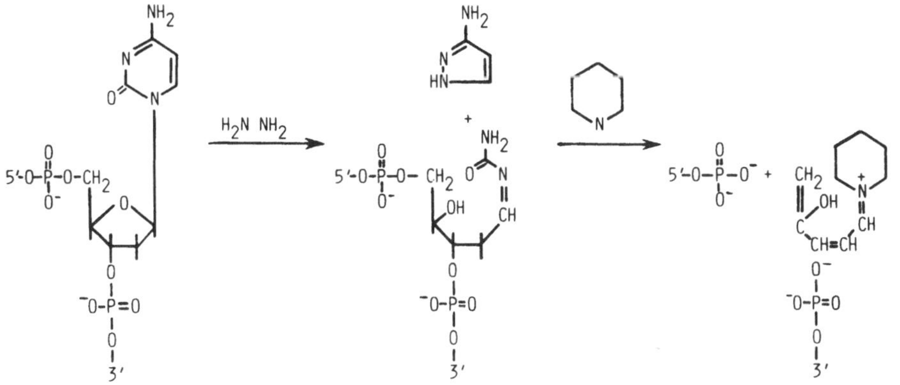

Figure A.3. Action of hydrazine on DNA. The limited reaction of hydrazine leads to the cleavage of DNA at pyrimidine bases and is made use of in the Maxam-Gilbert sequencing method.

point that, in high salt, when a methylcytosine is present in the DNA, hydrazine fails to act. This leads to a gap in the cytosine ladder wherever a methylcytosine occurs. At this point the complementary strand shows a guanine. This would appear to be the ultimate method for analysis of the location of methylcytosine, as every single modified base can be detected (not only those in CCGG sequences), and they can be precisely located in specific genes. However, there are two disadvantages. First, quantitation is difficult, and in general bands are not absent but only present at reduced intensity. Second, sequencing is primarily carried out on DNA cloned in *E. coli,* and hence it has the methylation pattern imparted not by its original cell but by *E. coli.* This second limitation does not apply to the sequencing of repetitive DNA fractions (*e.g.,* satellite DNA and ribosomal DNA), which can be isolated in pure form without cloning, nor to viral DNA's.

This second problem has been largely overcome by a modification of the method recently introduced by Church and Gilbert (1984) and shown in Figure A.4. Total genomic DNA is completely cleaved with a specific restriction enzyme and then subjected to the treatment described above with hydrazine and piperidine to produce one cleavage approximately every 500 nucleotides. The chance of cleavage at any particular cytosine is, therefore, small but even this low level of cleavage is abolished if the cytosine is methylated. The fragments are separated by electrophoresis under denaturing conditions and transferred and cross-linked to nylon membranes. These are then probed with a short (160–200 nucleotide) single-stranded ^{32}P-labeled probe specific for one end of a restriction fragment. In this way, a ladder of fragments is lit up corresponding to the specific restriction fragment cleaved at random by the chemical reagents.

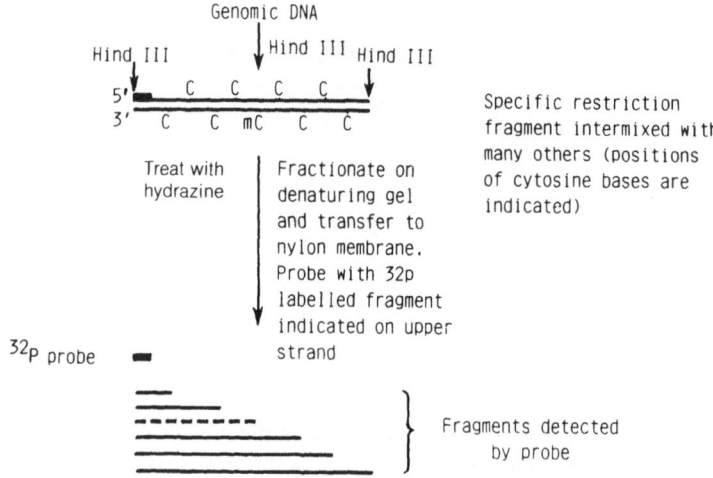

Figure A.4. Sequencing genomic DNA. The diagram illustrates the principle of the method of genomic sequencing introduced by Church and Gilbert (1984). Genomic DNA, cleaved with a restriction enzyme, is treated with hydrazine and the resulting digest is separated by electrophoresis on a sequencing gel. Following transfer to nylon membranes, the complex mixture is probed with a short radiolabeled probe. The diagram illustrates the main fragments that will be detected.

Where cleavage is blocked by methylation the intensity of the band is reduced. A number of potential problems are readily resolved (Church & Gilbert, 1984; Little, 1984), and as little as 3 fg can be detected in a single band. This means an analysis requires about 250 μg genomic DNA, but each membrane can be rehybridized to several probes.

As yet this method has not been extensively applied, but we might expect over the next few years that it will be the method of choice when it comes to locating methylcytosines in the DNA of various eukaryotic cells and tissues.

References

Abe, T., Okada, G., Teraoka, H. & Tsukada, K. (1982) J. Biochem. (Tokyo) *91*, 1081–1084.

Achwal, C.W., Iyer, C.A. & Chandra, H.S. (1983) FEBS Letts. *158*, 353–357.

Achwal, C.W., Ganguly, P. & Chandra, H.S. (1984) EMBO J. *3*, 263–266.

Adams, R.L.P. (1971) Biochim. Biophys. Acta *254*, 205–212.

———. (1973) Nature New Biology *244*, 27–29.

———. (1974) Biochim. Biophys. Acta *335*, 365–373.

———. (1980) Cell Culture for Biochemists. Elsevier/North Holland, Amsterdam.

Adams, R.L.P. & Burdon, R.H. (1982) CRC Crit. Rev. Biochem. *13*, 349–383.

———. (1983) In Enzymes of Nucleic Acid Synthesis and Modification, vol. 1:119–144, Jacob, S.T., ed. CRC Press Inc., Boca Raton, FL.

Adams, R.L.P. & Eason, R.E. (1984) Nucl. Acids Res. *12*, 5869–5877.

Adams, R.L.P. & Hogarth, C. (1973) Biochim. Biophys. Acta *331*, 214–220.

Adams, R.L.P., Turnbull, J., Smillie, E.J. & Burdon, R.H. (1974) Symposium FEBS, Budapest *34*, 39–48.

Adams, R.L.P., McKay, E.L., Douglas, J.T. & Burdon, R.H. (1977) Nucl. Acids Res. *4*, 3097–3108.

Adams, R.L.P., McKay, E.L., Craig, L.M. & Burdon, R.H. (1979a) Biochim. Biophys. Acta *561*, 345–357.

———. (1979b) Biochim. Biophys. Acta *563*, 72–81.

Adams, R.L.P., Burdon, R.H., Campbell, A.M., Leader, D.P. & Smellie, R.M.S. (1981a) The Biochemistry of the Nucleic Acids, 9th ed. Chapman & Hall, London.

Adams, R.L.P., Burdon, R.H., Gibb, S. & McKay, E.L. (1981b) Biochim. Biophys. Acta *655*, 329–334.

Adams, R.L.P., Fulton, J. & Kirk, D. (1982) Biochim. Biophys. Acta *697*, 286–294.

Adams, R.L.P., Burdon, R.H. & Fulton, J. (1983a) Biochem. Biophys. Res. Comm. *113*, 695–702.

Adams, R.L.P., Burdon, R.H., McKinnon, K. & Rinaldi, A. (1983b) FEBS Letts. *163*, 194–198.

Adams, R.L.P., Davis, T., Fulton, J., Kirk, D., Qureshi, M. & Burdon, R.H. (1984) Curr. Top. Microbiol. and Immunol. *108*, 143–156.

Adley, C.C. & Bukhari, A.I. (1984) Nucl. Acids Res. *12*, 3535–3550.

Akira, S., Sugiyama, H., Sakaguchi, N. & Kishimoto, T. (1984) EMBO J. *3*, 677–681.

Altanerova, V. & Altaner, C. (1972) Neoplasma *19*, 405–412.

Ammerman, D., Steinbruck, G., Bauer, R. & Wohlert, H. (1981) Eur. J. Cell Biol. *24*, 154–156.

Anderson, R.A., Kato, S. & Camerini-Otero, R.D. (1984) Proc. Natl. Acad. Sci. USA *81*, 206–210.

Andres, A-C., Muellener, D.B. & Ryffel, G.U. (1984) Nucl. Acids Res. *12*, 2283–2302.

Andrews, G.K., Dziadek, M. & Tamaoki, T. (1982) J. Biol. Chem. *257*, 5148–5153.

Antequera, F., Tamane, M., Villaneuva, J.R. & Santos, T. (1984) J. Biol. Chem. *259*, 8033–8036.

Antonov, A.S., Favorova, O.O. & Belozerksii, A.N. (1962) Dokl. Acad. Nauk, S.S.S.R. *147*, 1480–1483.

Arber, W. & Morse, M.L. (1965) Genetics *51*, 137–148.

Arnott, S., Chandarsekaran, R., Puigjaner, L.C., Walker, J.K., Hall, I.H., Birdsall, D.L. & Ratliff, R.L. (1983) Nucl. Acids Res. *11*, 1457–1474.

Arnt-Jovin, D.J., Robert-Nicoud, M., Zarling, D.A., Greider, C., Weiner, E. & Jovin, J.M. (1983) Proc. Natl. Acad. Sci. USA *80*, 4344–4348.

Atkinson, D.E. (1977) Cellular Energy Metabolism and its Regulation. Academic Press, New York. pp. 75.

Azorin, F., Nordheim, A. & Rich, A. (1983) EMBO J. *2*, 649–655.

Baldessarini, R.J. (1975) Int. Rev. Neurobiol. *18*, 41–67.

Baldessarini, R.J. & Kopin, I.J. (1966) J. Neurochem. *13*, 769.

Bale, A., d'Alarcao, M. & Marinus, M.G. (1979) Mutation Res. *59*, 157–165.

Ball, D.J., Gross, D.S. & Garrard, W.T. (1983) Proc. Natl. Acad. Sci. USA *80*, 5490–5494.

Barker, D., Schafer, M. & White, R. (1984) Cell *36*, 131–138.

Barrows, L.R. & Magee, P.N. (1982) Carcinogenisis *2*, 349–351.

Battistuzzi, G., D'Urso, M., Toniolo, D., Persico, G.M. & Luzzatto, L. (1985) Proc. Natl. Acad. Sci. USA *82*, 1465–1469.

Baur, R. & Kroger, H. (1976) Hoppe Seyler's Z. Physiol. Chem. *357*, 308.

Baur, R., Wohlert, H., Kluge, N. & Kroger, H. (1978a) 12th FEBS Meeting, Dresden, Abst. no. 642.

Baur, R., Wohlert, H. & Kroger, H. (1978b) Gesellschaft fur Biologische Chemie *359*, 254.

Baxter, R. & Kirk, J.T.O. (1969) Nature *222*, 272–273.

Behe, M. & Felsenfeld, G. (1981) Proc. Natl. Acad. Sci. USA *78*, 1619–1623.

Bendig, M.M. & Williams, J.G. (1983) Proc. Natl. Acad. Sci. USA *80*, 6197–6201.

Berdyshev, G.D., Korotaev, G.K., Boyarskikh, G.V. & Vanyushin, B.F. (1967) Biokhimiya *32*, 988–993.

Berg, O.G., Winter, R.B. & von Hippel, P.H. (1982) Trends in Biochem. Sci. *7*, 52–55.

Bernards, A., Harten-Loosbroek, van N., & Borst, P. (1984) Nucl. Acids Res. *12*, 4153–4170.

Bestor, T.H. & Ingram, V.M. (1983) Proc. Natl. Acad. Sci. USA *80*, 5559–5563.

Bickle, T.A., Brack, C. & Yuan, R. (1978) Proc. Natl. Acad. Sci. USA *75*, 3099–3103.

Billen, D. (1968) J. Mol. Biol. *31*, 477–486.

Bingham, A.H.A. & Atkinson, R. (1978) Biochem. Soc. Trans. *6*, 315–324.

Bird, A.P. (1977) Cold Spring Harbor Symp. Quant. Biol. *42*, 1179–1183.

——. (1978) J. Mol. Biol. *118*, 49–60.

——. (1980) Nucl. Acids Res. *8*, 1499–1504.

——. (1984a) Nature *307*, 503–504.

——. (1984b) Curr. Top. Microbiol. and Immunol. *108*, 129–141.

Bird, A.P. & Southern, E.M. (1978) J. Mol. Biol. *118*, 27–47.

Bird, A.P. & Taggart, M.H. (1980) Nucl. Acids Res. *8*, 1485–1497.

Bird, A., Taggart, M., Frommer, M., Miller, O.J. & Macleod, D. (1985) Cell *40*, 91–99.

Bird, A.P., Taggart, M.H. & Smith, B.A. (1979) Cell *17*, 889–901.

Bird, A.P., Taggart, M.H. & Gehring, C. (1981) J. Mol. Biol. *152*, 1–17.

Bird, A.P. Taggart, M.H. & Macleod, D. (1981) Cell *26*, 381–390.

Blackburn, E.H., Pan, W-C. & Johnson, C.C. (1983) Nucl. Acids Res. *11*, 5131–5145.

Blau, H.M., Chiu, C-P. & Webster, C. (1983) Cell *32*, 1171–1180.

Boehm, T.L. & Drahovsky, D. (1981a) Carcinogenisis *2*, 39–42.

——. (1981b) Int. J. Biochem. *13*, 1225–1232.

——. (1983) J. Natl. Cancer Inst. *71*, 429–433.

Bolden, A., Ward, C., Siedlecki, J.A. & Weissbach, A. (1984) J. Biol. Chem. *259*, 12437–12443.

Bolen, P.L., Grant, D.M., Swinton, D., Boynton, J.C. & Gilham, N.W. (1982) Cell *28*, 335–343.

Bonen, L., Huh, T.Y. & Gray, M.W. (1980) FEBS Letts. *111*, 340–346.

Borek, E. & Srinivasan, P.R. (1966) Ann. Rev. Biochem. *35*, 275.

Bostock, C. (1980) Trends in Biochem. Sci. *5*, 117–119.

Bougueleret, L., Schwarzstein, M., Tsugita, A. & Zabeau, M. (1984) Nucl. Acids Res. *12*, 3659–3676.

Bower, D.J., Errington, L.H., Cooper, D.N., Morrison, S. & Clayton, R.M. (1983) Nucl. Acids Res. *11*, 2513–2527.

Boyer, H.W. (1971) Ann. Rev. Microbiol. *25*, 153–176.

Boyer, H.W., Chow, L.T., Dugaiczyk, A., Hedgpath, J. & Goodman, H.M. (1973) Nature New Biology *244*, 40–43.

Brawerman, E. & Eisenstadt, J.M. (1964) Biochim. Biophys. Acta *91*, 477–485.

Brawerman, E., Hufnagel, D.A. & Chargaff, E. (1962) Biochim. Biophys. Acta *61*, 340–345.

Breathnach, R., Benoist, C., O'Hare, K., Gannon, F. & Chambon, P. (1978) Proc. Natl. Acad. Sci. USA *75*, 4853–4857.

Breznik, T. & Cohen, J.C. (1982) Nature *295*, 255–257.

Brinster, R.L., Ritchie, K.A., Hammer, R.E., O'Brien, R.L., Arp, B. & Storb, U. (1983) Nature *306*, 332–336.

Brinster, R.L., Chen, H.Y., Messing, A., van Dyke, T., Levine, A.J. & Palmiter, R.D. (1984) Cell *37*, 367–379.

Brockes, J.P., Brown, P.R. & Murray, K. (1972) Biochem. J. *127*, 1–10.

Brooks, J.E. & Roberts, R.J. (1982) Nucl. Acids Res. *10*, 913–934.

Brooks, J.E., Blumenthal, R.M. & Gingeras, T.R. (1983) Nucl. Acids Res. *11*, 837–851.

Brown, D.D. (1984) Cell *37*, 359–365.

Brown, D.D. & Dawid, I.B. (1968) Science *160*, 272–280.

Brown, D.D. & Gurdon, J.B. (1977) Proc. Natl. Acad. Sci. USA *74*, 2064–2068.

Brown, F.L., Musich, P.R. & Maio, J.J. (1978) Nucl. Acid Res. *5*, 1093–1107.

Browne, M.J. & Burdon, R.H. (1977) Nucl. Acids Res. *4*, 1025–1037.

Bugler, B., Bertaux, O. & Valencia, R. (1980) J. Cell Physiol. *104*, 149–157.

Bull, J.C. & Wootton, J.C. (1984) Nature *310*, 701–704.

Burch, J.B.E. & Weintraub, H. (1983) Cell *33*, 65–76.

Burckhardt, J., Weisemann, J. & Yuan, R. (1981) J. Biol. Chem. *256*, 4024–4032.

Burdon, R.H. (1966) Nature *210*, 797–799.

Burdon, R.H. & Adams, R.L.P. (1969) Biochim. Biophys. Acta *174*, 322–329.

———. (1980) Trends in Biochem. Sci. *5*, 294–297.

Burdon, R.H., Qureshi, M. & Adams, R.L.P. (1985) Biochim. Biophys. Acta *825*, 70–79.

Burton, W.G., Roberts, R.J., Myers, P.A. & Sager, R. (1977) Proc. Natl. Acad. Sci. USA *74*, 2687–2691.

Burton, W.G., Grabowy, C.T. & Sager, R. (1979) Proc. Natl. Acad. Sci. USA *76*, 1390–1394.

Buryanov, Y.I., Zinoviev, V.V., Vienozhinskis, M.T., Malygin, E.G., Nesterenko, V.F., Popov, S.G. & Gorbunov, Y.A. (1984) FEBS Letts. *168*, 166–168.

Busslinger, M. & Flavell, R.H. (1983) In Globin Gene Expression and Haemopoietic Differentiation, Nienhuis, A.W. & Stamatogannopoulos, G., eds. A.R. Liss, New York.

Busslinger, M., deBoer, E., Wright, S., Grosveld, F.G. & Flavell, R.A. (1983) Nucl. Acids Res. *11*, 3559–3569.

Busslinger, M., Hurst, J. & Flavell, R.A. (1983) Cell *34*, 197–206.

Calos, M.P., Lebkowski, J.S. & Botchan, M.R. (1983) Proc. Natl. Acad. Sci. USA *80*, 3015–3019.

Cantoni, G.L. (1960) In Comparative Biochemistry, 181–241, Florkin, M. & Mason, H.S., eds. Academic Press, New York.

———. (1975) Ann. Rev. Biochem. *44*, 435–451.

Cantoni, G.L. & Davies, D.R. (eds) (1966) Procedures in Nucleic Acid Research. Harper & Row, New York.

Cantoni, G.L. & Durell, J. (1957) J. Biol. Chem. 225, 1033–1048.

Cantoni, G.L. & Scarano, E. (1954) J. Amer. Chem. Soc. *76*, 4744.

Carrasco, A.E., McGinnis, W., Gehring, W.J. & de Robertis, E.M. (1984) Cell *37*, 409–414.

Cassidy, P., Kahan, F. & Alegria, A.A. (1965) Fed. Proc. *24*, 2I, 226.

Cate, R.L., Chick, W. & Gilbert, W. (1983) J. Biol. Chem. *258*, 6645–6652.

Cato, A.C.B., Adams, R.L.P. & Burdon, R.H. (1978) Biochim. Biophys. Acta *521*, 397–406.

Cattanach, B.M. (1975) Ann. Rev. Genet. *9*, 1–18.

Cedar, H., Solage, A., Glaser, G. & Razin, A. (1979) Nucl. Acids Res. *6*, 2125–2132.

Cedar, H., Stein, R., Gruenbaum, Y., Naveh-Many, T., Sciaky-Gallili, N. & Razin, A. (1983) Cold Spring Harbor Symp. Quant. Biol. *47*, 605–609.

Chambers, J.C. & Taylor, J.H. (1982) Chromosoma *85*, 603–609.

Chang, S. & Cohen, S.N. (1977) Proc. Natl. Acad. Sci. USA *74*, 4811–4815.

Chao, M.V., Mellon, P., Charney, P., Maniatis, T. & Axel, R. (1983) Cell *32*, 483–493.

Chapman, V.M., Kratzer, P.G., Siracusa, L.D., Quarantillo, B.A., Evans, R. & Liskay, R.M. (1982) Proc. Natl. Acad. Sci. USA *79*, 5357–5361.

Chapman, V., Forrester, L., Sanford, J., Hastie, N. & Rossant, J. (1984) Nature *307*, 284–286.

Charache, S., Dover, G.J., Smith, K.D., Talbot, C.C., Moyer, M. & Boyer, S. (1983) Proc. Natl. Acad. Sci. USA *80*, 4842–4846.

Chargaff, E. (1955) The Nucleic Acids Vol *1*, 307–371, Chargaff, E. & Davidson, J.N. eds. Academic Press, New York.

Chargaff, E., Lipshitz, R. & Green, C. (1952) J. Biol. Chem. *195*, 155–160.

Chen, M-J. & Nienhuis, A.W. (1981) J. Biol. Chem. *256*, 9680–9681.

Cheng, S-C., Herman C. & Modrich, P. (1985) J. Biol. Chem. *260*, 191–194.

Chiang, P.K. & Cantoni, G.L. (1977) J. Biol. Chem. *252*, 4506–4513.

Chiu, C-P. & Blau, H.M. (1984) Cell *37*, 879–887.

Chou, T-C., Coulter, A.W., Lombardini, J.B., Sufrin, J.R. & Talalay, P. (1977) In The Biochemistry of Adenosylmethionine, 18–36, Salvatore, F., Borek, E., Zappia, V., Williams-Ashman, H.G. & Schlenk, F., eds. Columbia University Press, New York.

Christman, J.K. (1984) Curr. Top. Microbiol. and Immunol. *108*, 49–78.

Christman, J.K., Price, P., Pedrinan, L. & Acs, G. (1977) Eur. J. Biochem. *81*, 53–61.

Christman, J.K., Schneiderman, N. & Aes, G. (1985) J. Biol. Chem. *260*, 4059–4068.

Christy, B. & Scangos, G. (1982) Proc. Natl. Acad. Sci. USA *79*, 6299–6303.

———. (1984) Mol. Cell Biol. *4*, 611–617.

Church, G.M. & Gilbert, W. (1984) Proc. Natl. Acad. Sci. USA *81*, 1991–1995.

Cihak, A. & Broucek, J. (1972) Biochem. Pharmacol. *21*, 2497–2507.

Cihak, A. & Kroger, H. (1981) Biomedicine *34*, 18–22.

Cihak, A. & Vesely, J. (1972) Biochem. Pharmacol. *21*, 3257–3265.

———. (1973) J. Biol. Chem. *248*, 1307–1313.

———. (1977) FEBS Letts. *78*, 244–246.

Cihak, A., Vesely, J., Inoue, H. & Pitot, H.C. (1972) Biochem. Pharmacol. *21*, 2545–2553.

Cihak, A., Seifertova, M., Vesely, J. & Sorm, F. (1972) Int. J. Cancer *10*, 20–27.

Cihak, A., Narurkar, L.M. & Pitot, H.C. (1973) Collect. Czech. Chem. Comm. *38*, 948–955.

Ciomei, M., Spadari, S., Pedrali-Noy, G. & Ciarrocchi, G. (1984) Nucl. Acids Res. *12*, 1977–1989.

Clough, D.W., Kunkel, L.M. & Davidson, R.L. (1982) Science *216*, 70–73.

Clutterbuck, A.J. & Spathas, D.H. (1984) Genet. Res. *43*, 123–138.

Cohen, J.C. (1980) Cell *19*, 653–662.

Cohen, S.S. & Barner, H.D. (1957) J. Biol. Chem. *226*, 631–642.

Comings, D.E. (1972) Exptl. Cell Res. *74*, 383–390.

Compere, S.J. & Palmiter, R.D. (1981) Cell *25*, 233–240.

Constantinides, P.C., Taylor, S.M. & Jones, P.A. (1978) Dev. Biol. *66*, 57–71.

Contreras, R., Gheyson, D., Knowland, J., van de Voorde, A. & Fiers, W. (1982) Nature *300*, 500–505.

Cooney, C.A., Mathews, H.R. & Bradbury, E.M. (1984) Nucl. Acids Res. *12*, 1501–1515.

Cooper, D.N. (1983) Hum. Genet. *64*, 315–333.

Cooper, D.N., Taggart, M.II. & Bird, A.P. (1983) Nucl. Acids Res *11*, 647–658.

Coulondre, C., Miller, J.H., Farabaugh, P.J. & Gilbert, W. (1978) Nature *274*, 775–780.

Coulter, A.W., Lombardini, J.B., Suffrin, R. & Talalay, P. (1974) Mol. Pharmacol. *10*, 319–334.

Cox, R. (1980) Cancer Res. *40*, 61–63.

Cox, R. & Irving, C.C. (1977) Cancer Res. *37*, 222–225.

Cox, R., Prescott, C. & Irving, C.C. (1977) Biochim. Biophys. Acta *474*, 493–499.

Creusot, F. & Christman, J.K. (1981) Nucl. Acids Res. *9*, 5359–5381.

Creusot, F., Acs, G. & Christman, J.K. (1982) J. Biol. Chem. *257*, 2041–2048.

Culp, L.A. & Black, P.H. (1971) Biochim. Biophys. Acta *247*, 220–232.

Cummings, D.J., Tait, A. & Goddard, J.M. (1974) Biochim. Biophys. Acta *374*, 1–11.

Darmon, N., Nicolas, J-F. & Lamblin, D. (1984) EMBO J. *3*, 961–967.

Davie, J.R. & Saunders, C.A. (1981) J. Biol. Chem. *256*, 12574–12580.

Davies, R.L., Fuhrer-Krusi, S. & Kucherlapati, R.S. (1982) Cell *31*, 521–529.

Davis, T., Kirk, D., Rinaldi, A., Burdon, R.H. & Adams, R.L.P. (1985) Biochem. Biophys. Res. Comm. *126*, 678–684.

Dawid, I.B. (1965) J. Mol. Biol. *12*, 581–599.

———. (1974) Science *184*, 80–81.

Dawid, I.B., Brown, D.D. & Reeder, R.H. (1970) J. Mol. Biol. *51*, 341–360.

Deguchi, T. & Barchas, J. (1971) J. Biol. Chem. *246*, 3175–3181.

de Jonge, A.J.R., Abrahams, P.J., Westerveld, A. & Bootsma, D. (1982) Nature *295*, 624–626.

De la Haba, G. & Cantoni, G.L. (1959) J. Biol. Chem. *234*, 603–608.

Deobagkar, D.N., Muralidharan, K., Devare, S.G., Kalghatgi, K.K. & Chandra, H.S. (1982) J. Biosci. *4*, 513–526.

Desai, L.S., Wulff, U.C. & Foley, G.E. (1971) Exptl. Cell Res. *65*, 260–263.

Deschatrette, J. (1980) Cell *22*, 501–511.

DeSimone, J., Heller, P., Hall, L. & Zuriers, D. (1982) Proc. Natl. Acad. Sci. USA *79*, 4428–4431.

Desrosiers, R.C., Mulder, C. & Fleckenstein, B. (1979) Proc. Natl. Acad. Sci. USA *76*, 3839–3843.

Deumling, B. (1981) Proc. Natl. Acad. Sci. USA *78*, 338–342.

Deuring, R. & Doerfler, W. (1983) Gene *26*, 283–289.

Deutsch, J., Razin, A. & Sedat, J.W. (1976) Anal. Biochem. *72*, 586–592.

Diala, E.S. & Hoffman, R.M. (1982) Biochem. Biophys. Res. Comm. *104*, 1489–1494.

Diala, E.S. & Hoffman, R.M. (1983) J. Virol. *45*, 482–483.

Dickerson, R.E. (1983a) Scientific American *249* (6), 86–103.

———. (1983b) In Nucleics Acids: The Vectors of Life, 1–15, Pullman, B. & Jortner, J., eds. D. Reidel, Dordrecht, Holland.

Dingwall, C., Lomonossoff, G.P. & Laskey, R.A. (1981) Nucl. Acids Res. *9*, 2659–2673.

Doerfler, W. (1981) J. Gen. Virol. *57*, 1–20.

———. (1983) Annu. Rev. Biochem. *52*, 93–124.

———. (1984) Current Topics in Microbiol and Immunol. *108*, 79–98.

Duoerfler, W., Kruczek, I., Eick, D., Vardimon, L. & Kron, B. (1978) Cold Spring Harbor Symp. Quant. Biol. *47*, 593–602.

Doskocil, J. & Sorm, F. (1962) Biochim. Biophys. Acta *55*, 953–959.

Drahovsky, D. & Boehm, T.L.J. (1980) Int. J. Biochem. *12*, 523–528.

Drahovsky, D. & Morris, N.R. (1971) J. Mol. Biol. *57*, 475–489.

Drahovsky, D. & Wacker, A. (1975) Naturewissenschaften *62*, 189.

Drahovsky, D., Lacko, I. & Wacker, A. (1976) Biochim. Biophys. Acta *447*, 139–143.

Drahovsky, D., Boehm, T.L.J. & Kreis, W. (1979) Biochim. Biophys. Acta *563*, 28–35.

Drahovsky, D., Kaul, S., Boehm, T.L.J. & Wacker, A. (1980) Biochim. Biophys. Acta *607*, 201–205.

Duerre, J.A. & Walker, R.D. (1977) In The Biochemistry of Adenosylmethionine, 43–57, Salvatore, F., Borek, E., Zappia, V., Williams-Ashman, H.G. & Schlenk, F., eds. Columbia University Press, New York.

Dybvig, K., Swinton, D., Maniloff, J. & Hattman, S. (1982) J. Bacteriol. *151*, 1420–1424.

Eastman, E.M., Goodman, R.M., Erlanger, B.F. & Miller, O.J. (1980) Chromosoma (Berlin) *79*, 225–239.

Efstratiadis, A., Posakony, J.W., Maniatis, T., Lawn, R.H., O'Connell, C., Sprintz, R.A., De Riel, J.K., Forget, B.G., Weissman, S.M., Slighton, J.L., Blechl, A.E., Smithies, O., Baralle, F.E., Shoulders, C.C. & Proudfoot, N.J. (1980) Cell *21*, 653–668.

Ehrlich, M. & Wang, R.Y-H. (1981) Science *212*, 1350–1357.

Ehrlich, M., Ehrlich, K. & Mayo, J.A. (1975) Biochim. Biophys. Acta *395*, 109–119.

Ehrlich, M., Gama-Sosa, M.A., Huang, L-H., Midgett, R.H., Kuo, K.C., McCune, R.A. & Gehrke, C. (1982) Nucl. Acids Res. *10*, 2709–2721.

Eick, D., Fritz, H-J., & Doerfler, W. (1983) Anal. Biochem. *135*, 165–171.

Ellis, T.H.N., Goldsbrough, P.B. & Castleton, J.A. (1983) Nucl. Acids Res. *11*, 3047–3064.

Engel, J.D. & von Hippel, P.H. (1974) Biochemistry *13*, 4143–4158.

——. (1978) J. Biol. Chem. *253*, 927–934.

Eskin, B. & Linn, S. (1972) J. Biol. Chem. *247*, 6183–6191.

Evans, M.J. (1972) J. Embryol. Exp. Morphol. *28*, 163.

Evans, H.H. & Evans, T.E. (1970) J. Biol. Chem. *245*, 6436–6441.

Evans, H.H., Evans, T.E. & Littman, S. (1973) J. Mol. Biol. *74*, 563–572.

Fabricant, J.D., Wagner, E.F., Auer, B. & Schweiger, M. (1979) Exptl. Cell Res. *124*, 25–29.

Fasman, G. (1976) Handbook of Biochemistry and Molecular Biology, 3d ed., vol. 2, 241–283. CRC Press Inc., Cleveland, OH.

Feher, Z., Kiss, A., & Venetianer, P. (1983) Nature *302*, 266–268.

Feinberg, A.P. & Vogelstein, B. (1983a) Nature *301*, 89–92.

——. (1983b) Biochem. Biophys. Res. Comm. *111*, 47–54.

Feinstein, S.C., Ross, S.R. & Yamamoto, K.R. (1982) J. Mol. Biol. *156*, 549–565.

Feinstein, S.I., Miller, D.A., Ehrlich, M., Gehrke, C.W., Eden, L.B. & Miller, O.J. (1985a) Biochim. Biophys. Acta *824*, 336–340.

Feinstein, S.I., Racaniello, V.R., Ehrlich, M., Gehrke, C.W., Miller, D.A. & Miller, O.J. (1985b) Nucl. Acids. Res. *13*, 3969–3978.

Felsenfeld, G. & McGhee, J. (1982) Nature *296*, 602–603.

Feng, T-Y. & Chiang, K-S. (1984) Proc. Natl. Acad. Sci. USA *81*, 3438–3442.

Fiedler, W., Nobis, P., Jahner, D. & Jaenisch, R. (1982) Proc. Natl. Acad. Sci. USA *79*, 1874–1878.

Fisher, E.F. & Caruthers, M.H. (1979) Nucl. Acids Res. *7*, 401–416.

Folger, K., Anderson, J.N., Hayward, M.A. & Shapiro, D.J. (1983) J. Biol. Chem. *258*, 8908–8914.

Ford, J.P., Coca-Prados, M. & Hsu, M-T. (1980) J. Biol. Chem. *255*, 7544–7547.

Fradin, A., Manley, J.L. & Prives, C.L. (1982) Proc. Natl. Acad. Sci. USA *79*, 5142–5146.

Fraser, N.W., Burdon, R.H. & Elton, R.A. (1975) Nucl. Acids Res. *2*, 2131–2146.

Friedman, J. & Razin, A. (1976) Nucl. Acids Res. *3*, 2665–2675.

Friedman, J., Friedman, A. & Razin, A. (1977) Nucl. Acids Res. *4*, 3483–3496.

Fry, K., Poon, R., Whitcome, P., Idriss, J., Salser, W., Mazrimas, J. & Hatch, R. (1973) Proc. Natl. Acad. Sci. USA *70*, 2642–2646.

Fuchs, R. & Danne, M. (1972) Biochemistry *11*, 2659–2666.

Fujimura, F.K., Deininger, P.L., Friedmann, T. & Linney, E. (1981) Cell *23*, 809–814.

Fuller, R.S. & Kornberg, A. (1983) Proc. Natl. Acad. Sci. USA *88*, 5817–5821.

Furlong, J.C. & Maden, B.E.H. (1983) EMBO J. *2*, 443–448.

Gall, J.G. (1974) Proc. Natl. Acad. Sci. USA *71*, 3078–3081.

Gama-Sosa, M.A., Wang, R.Y-H., Kuo, K.C., Gehrke, C.W. & Ehrlich, M. (1983) Nucl. Acids Res. *11*, 3087–3095.

Gama-Sosa, M.A., Slagel, V.A., Trewyn, R.W., Oxenhandler, R., Kuo, K.C., Gehrke, C.W. & Ehrlich, M. (1983) Nucl. Acids Res. *11*, 6883–3894.

Gama-Sosa, M.A., Midgett, R.M., Slagel, V.A., Githens, S., Kuo, K.C., Gehrke, C.W. & Ehrlich, M. (1983c) Biochim. Biophys. Acta *740*, 212–219.

Gates, F.T. & Linn, S. (1977) J. Biol. Chem. *252*, 1647–1653.

Gattoni, S., Kirschmeier, P., Weinstein, I.B., Escobedo, J. & Dina, D. (1982) Mol. Cell Biol. *2*, 42–51.

Gautsch, J.W. (1980) Nature *285*, 110–112.

Gautsch, J.W. & Wilson, M.C. (1983) Nature *301*, 32–37.

Geier, G.E. & Modrich, P. (1979). J. Biol. Chem. *254*, 1408–1413.

Gelvin, S.B., Karcher, S.J. & DiRita, V.S. (1983) Nucl. Acids Res. *11*, 159–174.

Geraci, D., Eremenko, T., Cocchiara, R., Granieri, A., Scarano, E. & Volpe, P. (1974) Biochem. Biophys. Res. Comm. *57*, 353–358.

Gerber-Huber, S., May, F.E.B., Westley, B.R., Felber, B.K., Hosbach, H.A., Andres, A-C. & Ryfell, G.U. (1983) Cell *33*, 43–51.

German, D.C., Bloch, C.A. & Kredich, N.M. (1983) J. Biol. Chem. *258*, 10997–11003.

Gerondakis, S., Boyd, A., Bernard, O., Webb, E. & Adams, J.M. (1984) EMBO Journal *3*, 3013–3021.

Gill, J.E., Mazrimas, J.A. & Bishop, C.C. (1974) Biochim. Biophys. Acta *335*, 330–348.

Gjerset, R.A. & Martin, D.W. (1982) J. Biol. Chem. *257*, 8581–8583.

Glickman, B.W. & Radman, M. (1980) Proc. Natl. Acad. Sci. USA *27*, 1063–1067.

Glover, S.W. (1970) Genet. Res. *15*, 237–250.

Goddard, J.P. & Schulman, L.H. (1972) J. Biol. Chem. *247*, 3864–3867.

Gomez-Eichelmann, M.C. & Lark, K.G. (1977) J. Mol. Biol. *117*, 621–635.

Gorovsky, M.A., Hattman, S. & Pleger, G.L. (1973) J. Cell Biol. *56*, 697–701.

Graessmann, M., Graessmann, A., Wagner, H., Werner, E. & Simon, D. (1983) Proc. Natl. Acad. Sci. USA *80*, 6470–6474.

Grainger, R.M., Hazard-Leonards, R.M., Samaha, F., Hougan, L.M., Lesk, M.R. & Thomsen, G.H. (1983) Nature *306*, 88–91.

Graziani, F., Parisi, E. & Scarano, E. (1970) Biochim. Biophys. Acta *213*, 208–214.

Greenberg, D.M. (1963) Adv. Enzymol. *25*, 395–431.

Greene, P.J., Gupta, M., Boyer, H.W., Brown, W.E. & Rosenberg, J.M. (1981) J. Biol. Chem. *256*, 2143–2153.

Grippo, P., Iaccarino, M., Parisi, E. & Scarano, E. (1968) J. Mol. Biol. *36*, 195–208.

Groot, G.S.P. & Kroon, A.M. (1979) Biochim. Biophys. Acta *562*, 355–357.

Groudine, M. & Conkin, K.F. (1985) Science *228*, 1061–1068.

Groudine, M. & Weintraub, H. (1981) Cell *24*, 393–401.

Groudine, M., Eisenman, R. & Weintraub, H. (1981) Nature *292*, 311–317.

Gruenbaum, Y., Naveh-Many, T., Cedar, H. & Razin, A. (1981) Nature *292*, 860–862.

Gruenbaum, Y., Stein, R., Cedar, H. & Razin, A. (1981) FEBS Letts. *124*, 67–71.

Gruenbaum, Y., Cedar, H. & Razin, A. (1982) Nature *295*, 620–622.

Gruenbaum, Y., Szyf, M., Cedar, H. & Razin, A. (1983) Proc. Natl. Acad. Sci. USA *80*, 4919–4921.

Guha, S. (1984) Eur. J. Biochem. *145*, 99–106.

Gunthert, U. & Trautner, T.A. (1984) Curr. Top. Microbiol. and Immunol. *108*, 11–22.

Gunthert, U., Schwieger, M., Stupp, M. & Doerfler, W. (1976) Proc. Natl. Acad. Sci. USA *73*, 3923–3927.

Gunthert, U., Freund, M. & Trautner, T.A. (1981a) J. Biol. Chem. *256*, 9340–9345.

Gunthert, U., Jentsch, S. & Freund, M. (1981b) J. Biol. Chem. *256*, 9346–9351.

Gunzberg, W.H. & Groner, B. (1984) EMBO J. *3*, 1129–1135.

Guseinov, V.A., Mazin, A.L., Vanyushin, B.F. & Belozersky, A.N. (1972) Biokhimiya *37*, 312–318.

Guseinov, V.A., Kiryanov, G.I. & Vanyushin, B.F. (1975) Mol. Biol. Rep. *2*, 59–63.

Hadi, S.M., Bachi, B., Iida, S. & Bickle, T.A. (1983) J. Mol. Biol. *165*, 19–34.

Haigh, L.S., Owens, B.B., Hellewell, S. & Ingram, V.M. (1982) Proc. Natl. Acad. Sci. USA *79*, 5332–5336.

Hamada, H. & Kakunaga, T. (1982) Nature *298*, 396–398.

Hamada, H., Petrino, M.G. & Kakunaga, T. (1982) Proc. Natl. Acad. Sci. USA *79*, 6465–6469.

Harbers, K., Harbers, B. & Spencer, J.H. (1975) Biochem. Biophys. Res. Comm. *66*, 738–746.

Harbers, K., Jahner, D. & Jaenisch, R. (1981) Nature *293*, 540–542.

Harland, R.M. (1982) Proc. Natl. Acad. Sci. USA *79*, 2323–2327.

Harris, H. (1970) In Cell Fusion. Clarendon Press, Oxford.

Harrison, J.J., Anisowicz, A., Gadi, I.K., Raffeld, M. & Sager, R. (1983) Proc. Natl. Acad. Sci. USA *80*, 6606–6610.

Hattman, S., Kenny, C., Berger, L. & Pratt, K. (1978) J. Bacteriol. *135*, 1156–1157.

Hattman, S., Gribbin, C. & Hutchison, C.A. (1979) J. Virol. *32*, 845–851.

Hattman, S., Goradia, M., Monaghan, C. & Bukhari, A.I. (1983) Cold Spring Harbor Symp. Quant. Biol. *47*, 637–653.

Hecht, N.B., Liem, H., Kleene, K.C., Distel, R.J. & Ho, S-M. (1984) Dev. Biol. *102*, 452–461.

Helland, S. & Ueland, P.M. (1983) Cancer Res. *43*, 1847–1850.

Herbomel, P., Saragosti, S., Blangy, D. & Yaniv, M. (1981) Cell *25*, 651–658.

Herman, G.E. & Modrich, P. (1982) J. Biol. Chem. *257*, 2605–2612.

Hibasami, H., Hoffman, J.L. & Pegg, A.E. (1980) J. Biol. Chem. *255*, 6675–6678.

Hildersheim, J., Goguillan, J. & Lederer, E. (1973) FEBS Letts. *30*, 177–180.

Hill, R.J. & Stollar, B.D. (1983) Nature *305*, 338–340.

Hilliard, J.K. & Sneider, T.W. (1975) Nucl. Acids Res. *2*, 809–819.

Hjelle, B.L., Phillips, J.A. & Seeburg, P.H. (1982) Nucl. Acids Res. *10*, 3459–3474.

Hoekstra, M.F. & Malone, R.E. (1985) Molec. Cell. Biol. *5*, 610–618.

Hoffman, D.R., Marion, D.W., Cormatzer, W.E. & Duerre, J.A. (1980) J. Biol. Chem. *255*, 10822–10827.

Hoffman, J.L. & Kunz, G.L. (1980) Fed. Proc. Fed. Am. Soc. Exp. Biol. *39*, 1690.

Hoffman, J.L. & Sullivan, D.M. (1981) Fed. Proc. Fed. Am. Soc. Exp. Biol. *40*, 1832.

Hoffmann, J.W., Steffen, D., Gusella, J., Tabin, C., Bird, S., Cowing, D. & Weinberg, R.A. (1982) J. Virol. *44*, 144–157.

Holliday, R. & Pugh, J.E. (1975) Science *187*, 226–232.

Horiuchi, K. & Zinder, N.D. (1972) Proc. Natl. Acad. Sci. USA *69*, 3220–3224.

Hotta, Y. & Hecht, N. (1971) Biochim. Biophys. Acta *238*, 50–59.

Hu, W-S., Fanning, T.G. & Cardiff, R.D. (1984) J. Virol. *45*, 66–71.

Hughes, P., Squali-Houssaini, F-Z., Forterre, P. & Kohiyama, M. (1984) J. Mol. Biol. *176*, 155–159.

Hurwitz, J., Gold, M. & Anders, M. (1964) J. Biol. Chem. *239*, 3474–3482.

Iida, S., Meyer, J., Bachi, B., Stalhammar-Carlemalin, M., Schrickel, S., Bickle, T.A. & Arber, W. (1983) J. Mol. Biol. *165*, 1–18.

Illmensee, K. & Hoppe, P.C. (1981) Cell *23*, 9–18.

Ingram, R.S., Scott, R.W. & Tilghman, S.M. (1981) Proc. Natl. Acad. Sci. USA *78*, 4694–4698.

Jaenisch, R. & Jahner, D. (1984) Biochim. Biophys. Acta *782*, 1–9.

Jaenisch, R., Jahner, D., Nobis, P., Simon, I., Lohler, J., Harbers, K. & Grotkopp, D. (1981) Cell *24*, 519–529.

Jahner, D. & Jaenisch, R. (1985) Nature *315*, 594–597.

Jahner, D., Stuhlmann, H., Stewart, C.L., Harbers, K., Lohler, J., Simon, I. & Jaenisch, R. (1982) Nature *298*, 623–628.

Janulaitis, A., Klimasauskas, S., Petrusyte, M. & Butkus, V. (1983) FEBS Letts. *161*, 131–134.

Jolly, D.J., Okajama, H., Berg, P., Esty, A.C., Filpula, D., Bohlen, P., Johnson, G.G., Shively, J.E., Hunkapillar, T. & Friedman, T. (1983) Proc. Natl. Acad. Sci. USA *80*, 477–481.

Jones, P.A. & Taylor, S.M. (1980) Cell *20*, 85–93.

———. (1981) Nucl. Acids Res. *9*, 2933–2947.

Jones, P., Taylor, S.M., Mohandas, T. & Shapiro, L.J. (1982) Proc. Natl. Acad. Sci. USA *79*, 1215–1219.

Jove, R., Sperber, D.E. & Manley, J.L. (1984) Nucl. Acids Res. *12*, 4715–4730.

Kahmann, R. (1983) Cold Spring Harbor Symp. Quant. Biol. *47*, 639–646.

———. (1984) Curr. Top. Microbiol. and Immunol. *108*, 29–47.

Kallen, R.G., Simon, M. & Marmur, J. (1962) J. Mol. Biol. *5*, 248–250.

Kalousek, F. & Morris, N.R. (1969a) Science *164*, 721–722.

———. (1969b) J. Biol. Chem. *244*, 1157–1163.

Kalt, M.R. & Gall, J.G. (1974) J. Cell. Biol. *62*, 460–472.

Kan, N.C., Lautenberger, J.A., Edgell, M.H. & Hutchison, C.A. (1979) J. Mol. Biol. *130*, 191–209.

Kappler, J.W. (1970) J. Cell Physiol. *75*, 21–32.

———. (1971) J. Cell Physiol. *78*, 33–36.

Kaput, J. & Sneider, T.J. (1979) Nucl. Acids, Res. *7*, 2303–2322.

Karin, M., Haslinger, A., Holtgreve, H., Richards, R.I., Krauter, P., Westphal, H.L. & Beato, M. (1984) Nature *308*, 513–519.

Karon, M. & Benedict, W.F. (1972) Science, *178*, 62.

Karpov, V.L., Preobrazhenskaya, O.V. & Mirzabekov, A.D. (1984) Cell *36*, 423–431.

Kastan, M.B., Gowans, B.J. & Lieberman, M.W. (1982) Cell *30*, 509–516.

Kauc, L. & Piekarowicz, A. (1978) Eur. J. Biochem. *92*, 417–426.

Kaye, J.S., Bellard, M., Dretzen, G., Bellard, F. & Chambon, P. (1984) EMBO J. *3*, 1137–1144.

Keene, M.A. & Elgin, S.C.R. (1981) Cell *27*, 57–64.

Kerem, B-S., Goitein, R., Richler, C., Marcus, M. & Cedar, H. (1983) Nature *304*, 88–90.

Kerr, S.J. (1972) J. Biol. Chem. *247*, 4248–4252.

Kerr, S.J. & Borek, E. (1972) Adv. Enzymol. *36*, 1–27.

Keshet, E. & Cedar, H. (1983) Nucl. Acids Res. *11*, 3571–3580.

Kirk, J.T.O. (1967) J. Mol. Biol. *28*, 171–172.

Kiryanov, G.I., Kirnos, M.D., Demidkina, N.P., Alexandrushkina, N.I. & Vanyushin, B.F. (1980) FEBS Letts. *112*, 225–228.

Kiryanov, G.I., Isaeva, L.V., Kirnos, M.D., Ganicheva, N.I. & Vanyushin, D. (1982) Biochimica *47*, 132–139.

Klysik, J., Stirdivant, S.M. & Wells, R.D. (1982) J. Biol. Chem. *257*, 10152–10158.

Klysik, J., Stirdivant, S.M., Singleton, C.K., Zacharias, W. & Wells, R.D. (1983) J. Mol. Biol. *168*, 51–71.

Konieczny, S.F. & Emerson, C.P. (1984) Cell *38*, 791–800.

Korba, B.E. & Hays, J.B. (1982a) Cell *28*, 531–541.

———, (1982b) J. Mol. Biol. *157*, 213–235.

Kosykh, V.G., Buryanov, Y.I. & Bayev, A.A. (1980) Mol. Gen. Genet. *178*, 717–718.

Kramer, W., Schughart, K. & Fritz, H-J. (1982) Nucl. Acids Res. *10*, 6475–6485.

Kratzer, P.G., Chapman, V.M., Lambert, H., Evans, R.E. & Liskey, R.M. (1983) Cell *33*, 37–42.

Kredich, N.M. (1980) J. Biol. Chem. *255*, 7380–7385.

Kredich, N.M. & Hershfeld, M.S. (1979) Proc. Natl. Acad. Sci. USA *76*, 2450–2454.

Kredich, N.M. & Martin, D.W. (1977) Cell *12*, 931–938.

Kruczek, I. & Doerfler, W. (1982) EMBO J. *1*, 409–414.

———. (1983) Proc. Natl. Acad. Sci. USA *80*, 7586–7590.

Kuhlmann, I. & Doerfler, W. (1982) Virology *118*, 169–180.

Kunnath, L. & Locker, J. (1982a) Nucl. Acids Res. *10*, 3877–3892.

———. (1982b) Biochim. Biophys. Acta *699*, 264–271.

———. (1983) EMBO J. *2*, 317–324.

Kuo, K.C., McCune, R.A. & Gehrke, C.W. (1980) Nucl. Acids Res. *8*, 4763–4776.

Kuo, M.T., Mandel, J.L. & Chambon, P. (1979) Nucl. Acids Res. *7*, 2105–2113.

Kuo, T-T. & Tu, J. (1976) Nature *263*, 615.

Kuroiwa, T., Kawono, S. & Nishibayashi, S. (1982) Nature *298*, 481–483.

Lacks, S. & Greenberg, B. (1977) J. Mol. Biol. *114*, 153–168.

Lacy, E. & Axel, R. (1975) Proc. Natl. Acad. Sci. USA *72*, 3978–3982.

Lan, N.C. (1984) J. Biol. Chem. *259*, 11601–11606.

Langner, K-D., Vardimon, L., Renz, D. & Doerfler, W. (1984) Proc. Natl. Acad. Sci. USA *81*, 2950–2954.

Lapeyre, J.N. & Becker, F.F. (1979) Biochem. Biophys. Res. Comm. *37*, 698–705.

Lapeyre, J.N., Maizel, A.L. & Becker, F.F. (1980) Biochem. Biophys. Res. Comm. *95*, 630–637.

Lapeyre, J.N., Walker, M.S. & Becker, F.F. (1981) Carcinogenesis *2*, 873–878.

Lark, C. (1968a) J. Mol. Biol. *31*, 389–399.

——, (1968b) J. Mol. Biol *31*, 401–414.

Lautenberger, J.A. & Linn, S. (1972) J. Biol. Chem. *247*, 6176–6182.

Le Beau, M.M. & Rowley, J.D. (1984) Nature *308*, 607–608.

Lee, J.S., Woodsworth, M.L., Latimer, L.J.P. & Morgan, A.R. (1984) Nucl. Acids Res. *12*, 6603–6611.

Lee, T. & Karon, M.R. (1976) Biochem. Pharmacol. *25*, 1737–1742.

Lee, T., Karon, M. & Momparler, R.L. (1974) Cancer Res. *34*, 2482–2488.

Lennon, G.G. & Fraser, N.W. (1983) J. Mol. Evol. *19*, 286–288.

Levitan, I.B. & Webb, T.E. (1969) Biochim. Biophys. Acta *182*, 491–500.

Levy, W.P. & Walker, N.E. (1981) Biochemistry *20*, 1120–1127.

Ley, T.J., DeSimone, J., Anagnou, N.P., Keller, G.H., Humphries, R.K., Turner, P.H., Young, N.S., Heller, P. & Nienhuis, A.W. (1982) New Engl. J. Med. *307*, 1469–1475.

Ley, T.J., Chiang, Y.L., Haidaris, D., Anagnou, N.P., Wilson, V.L. & Anderson, W.F. (1984) Proc. Natl. Acad. Sci. USA *81*, 6618–6622.

Li, L.H., Olin, E.J., Fraser, T.J. & Bhuyan, B.K. (1970) Cancer Res. *30*, 2770–2775.

Lieberman, N.W., Beach, L.R. & Palmiter, R.D. (1983) Cell *35*, 207–214.

Lilley, D.M.J. (1983) Nature *305*, 276–277.

Lindahl, T., Ljungquist, S., Siegert, W., Nyberg, G. & Sperens, G. (1977) J. Biol. Chem. *252*, 3286–3294.

Linney, E., Davis, B., Overhauser, J., Chao, E. & Fou, H. (1984) Nature *308*, 470–472.

Lipps, H.J., Nordheim, A., Lofer, E.M., Ammermann, D., Stollar, B.D. & Rich, A. (1983) Cell *32*, 435–441.

Liskay, R.M. & Evans, R.J. (1980) Proc. Natl. Acad. Sci. USA *77*, 4895–4898.

Little, P.F.R. (1984) Nature *310*, 369.

Lombardini, J.B., Coulter, A.W. & Talalay, P. (1970) Mol. Pharmacol. *6*, 481–499.

Lombardini, J.B., Burch, M.K. & Talalay, P. (1971) J. Biol. Chem. *246*, 4465–4470.

Low, M., Hay, J. & Keir, H.M. (1969) J. Mol. Biol. *46*, 205–207.

Lu, A-L., Clark, S. & Modrich, P. (1983) Proc. Natl. Acad. Sci. USA *80*, 4639–4643.

Lu, L.W., Chiang, G.H., Medina, D. & Randerath, K. (1976) Biochem. Biophys. Res. Comm. *68*, 1094–1101.

Lutz, D., Lowel, M., Kroger, H., Kubsch, D. & Vecker, W. (1972) Zeit. Natur-forsch *27b*, 992–995.

Lutz, D., Grahn, H. & Kroger, H. (1973) Zeit. Naturforsch *28c*, 460–462.

Lyon, M.F. (1972) Biol. Rev. *47*, 1–35.

Macaya, G., Cortedas, J. & Bernardi, G. (1978) Eur. J. Biochem. *84*, 179–188.

Maden, B.E.H. & Salim, M. (1974) J. Mol. Biol. *88*, 133–164.

McClelland, M. (1981) Nucl. Acids Res. *9*, 5859–5866.

McClelland, M. & Ivarie, R. (1982) Nucl. Acids Res. *10*, 7865–7877.

McClelland, M. & Nelson, M. (1985) Nucl. Acids Res. *13 Suppl.*, r201–r207.

McGeady, M.L., Happan, C.J., Ascione, R. & van de Woude, G.F. (1983) Mol. Cell Biol. *3*, 305–314.

McGhee, J.D. & Ginder, G.D. (1979) Nature, *280*, 419–420.

McGhee, J.D., Wood, W.I., Dolan, M., Engel, J.D. & Felsenfeld, G. (1981) Cell *27*, 45–55.

McGinnis, W., Garber, R.L., Wirz, J., Kuroiwa, A. & Gehring, W.J. (1984) Cell *37*, 403–408.

McGrath, J. & Solter, D. (1983) Science *220*, 1300–1302.

McGraw, B.R. & Marinus, M.G. (1980) Mol. Gen. Genet. *178*, 309–315.

McKeon, C., Ohkubo, H., Pastan, I. & de Crombrugghe, B. (1982) Cell *29*, 203–210.

McKeon, C., Pastan, I. & de Crombrugghe, B. (1984) Nucl. Acids Res. *12*, 3491–3502.

McKnight, S.L., Kingsbury, R.C., Spence, A. & Smith, M. (1984) Cell *37*, 253–262.

Macleod, D. & Bird, A.P. (1982) Cell *29*, 211–218.

———. (1983) Nature *306*, 200–203.

Maeda, N., Yang, F., Barnett, D.R., Bowman, B.H. & Smithies, O. (1984) Nature *309*, 131–135.

Mamelak, L. & Boyer, H.W. (1970) J. Bact. *104*, 57–62.

Mandel, J.L. & Chambon, P. (1979) Nucl. Acids Res. *7*, 2081–2103.

Manes, C. & Menzel, P. (1981) Nature *293*, 589–590.

Maniatis, T., Fritsch, E.F. & Sambrook, J. (1982) Molecular Cloning p 458–460. Cold Spring Harbor Lab.

Manley, J.L., Fire, A., Cano, A., Sharp, P.A. & Gefter, M.L. (1980) Proc. Natl. Acad. Sci. USA *77*, 3855–3859.

Mann, M.B., Reo, R.N. & Smith, H.O. (1978) Gene *3*, 97–112.

Manton, S.M. & Anderson, D.T. (1979) in The Origin of Major Invertebrate Groups, 269–321, House, M.R., ed. Academic Press, New York.

Manuelidis, L. (1981) FEBS Letts. *129*, 25–28.

Marcaud, L., Reynaud, C-A., Therwath, A. & Scherrer, K. (1981) Nucl. Acids Res. *9*, 1841–1851.

Marinus, M.G. (1976) J. Bact. *128*, 853–854.

———. (1981) In Chromosome Damage and Repair, 469–473, Seeberg, E. & Kleppe, K., eds. Plenum, New York.

Marinus, M.G. & Morris, N.R. (1973) J. Bact. 114, 1143–1150.

———. (1974) J. Mol. Biol. *85*, 309–323.

Marinus, M.G., Poteete, A. & Arraj, J.A. (1984) Gene *28*, 123–125.

Mark, K-K. & Studier, F.W. (1981) J. Biol. Chem. *256*, 2573–2578.

Markham, G.D., Hafner, E.W., Tabor, C.W. & Tabor, H. (1980) J. Biol. Chem. *255*, 9082–9092.

Martin, G.R. (1982) Cell *29*, 721–724.

Martin, G.R. & Evans, M.J. (1975) Cell *6*, 467–474.

Mather, A.N. (1958) Fed. Proc. Fed. Am. Soc. Exp. Biol. *17*, 271.

Mather, E.L. & Perry, R.P. (1983) Proc. Natl. Acad. Sci. USA *80*, 4689–4693.

Matsumoto, L.H. (1981) Nature *294*, 481–482.

Max, E.E. (1984) Nature, *310*, 100.

Maxam, A. & Gilbert, W. (1977) Proc. Natl. Acad. Sci. USA *74*, 560–564.

———. (1980) Methods Enzymol. *65*, 499–560.

Mayo, K. & Palmiter, R.D. (1981) J. Biol. Chem. *256*, 2621–2624.

Meijer, M., Beck, E., Hansen, F.G., Bergmans, H.E.N., Messer, E., von Meyenberg, K. & Schaller, H. (1979) Proc. Natl. Acad. Sci. USA *76*, 580–584.

Meijlink, F.C.P.W., Philipsen, J.N.J., Gruber, M. & Geert, A.B. (1983) Nucl. Acids Res. *11*, 1361–1373.

Mermod, J-J., Bourgeois, S., Defer, N. & Crepin, M. (1983) Proc. Natl. Acad. Sci. USA *80*, 110–114.

Meselson, M., Yuan, R. & Heywood, J. (1972) Ann. Rev. Biochem. *41*, 447–466.

Michelson, A.M. (1963) in Chemistry of Nucleosides and Nucleotides, 45–60. Academic Press, New York.

Mikol, Y.B., Hoover, K.L., Creasia, D. & Poirier, L.A. (1983) Carcinogenesis *4*, 1619–1629.

Miller, D.A., Okamoto E., Erlanger, B.F. & Miller, O.J. (1982) Cytogenet. Cell Genet. *33*, 345–349.

Miller, O.J., Schnedl, W., Allen, J. & Erlanger, B.F. (1974) Nature *251*, 636–637.

Miller, O.J., Tantravahi, U., Katz, R., Erlanger, B.F. & Guntaka, R.V. (1981) In Genes, Chromosomes and Neoplasia, 253–270, Arrighi, F.E., Rao, P.N. & Stubblefield, E., eds. Raven Press, New York.

Miller, R.H. & Robinson, W.S. (1983) Proc. Natl. Acad. Sci. USA *80*, 2534–2538.

Mintz, B. & Illmensee, K. (1975) Proc. Natl. Acad. Sci. USA *72*, 3585–3589.

Modrich, P. & Roberts, R.J. (1982) In The Nucleases, 109–154, Cold Spring Harbor Laboratory Monograph no. 14 Linn, S.M. & Roberts, R.J., eds. Cold Spring Harbor, New York.

Mohandas, T., Sparkes, R.S. & Shapiro, L.J. (1981) Science *211*, 393–396.

Momparler, R.L. & Derse, D. (1979) Biochem. Pharmacol. *28*, 1443–1444.

Mondello, C. & Goodfellow, P.N. (1985) Trends in Genet. *1*, 124–125.

Monk, M. & Harper, M.I. (1979) Nature *281*, 311–313.

Morgenegg, G., Celio, M.R., Malfoy, B., Leng, M. & Kuenzle, C.C. (1983) Nature *303*, 540–543.

Morris, N.R. & Pih, K.D. (1971) Cancer Res. *31*, 433–440.

Mudd, S.H. (1959) J. Biol. Chem. *234*, 87–92.

———. (1963) J. Biol. Chem. *238*, 2156–2163.

———. (1965) In Transmethylation and Methionine Biosynthesis, 33, Shapiro, S.K. & Schlenk, F., eds. University of Chicago Press, Chicago.

Mudd, S.H. & Cantoni, G.L. (1957) Nature *180*, 1052.

———. (1958) J. Biol. Chem. *231*, 481–492.

———. (1964) in *Comprehensive Biochemistry* Vol 15, 1–27. Florkin, M. & Stotz, E.H. eds. Elsevier.

Mudd, S.H. & Levy, H.L. (1983) in Metabolic Basis of Inherited Diseases, 5th ed., 522–559, Stanbury, J.B., ed. Ed. 5 McGraw-Hill, New York.

Mudd, S.H. & Mann, J.D. (1963) J. Biol. Chem. *238*, 2164–2170.

Mudd, S.H., Ebert, M.H. & Scriver, C.R. (1980) Metabolism *29*, 707–720.

Nakhasi, H.L., Lynch, K.R., Dolan, K.P., Unterman, R.D. & Feigelson, P. (1981) Proc. Natl. Acad. Sci. USA *78*, 834–837.

Nardone, G., George, J. & Chirikjian, J.G. (1984) J. Biol. Chem. *259*, 10357–10362.

Nass, M.M.K. (1973) J. Mol. Biol. *80*, 155–175.

Naveh-Many, T. & Cedar, H. (1981) Proc. Natl. Acad. Sci. USA *78*, 4246–4250.

———. (1982) Mol. Cell Biol. *2*, 758–762.

Nelson, P.P., Albright, S.C. & Garrard, W.T. (1979) J. Biol. Chem. *254*, 9194–9199.

Newman, A.K., Rubin, R.A., Kim, S-H. & Modrich, P. (1981) J. Biol. Chem. *256*, 2131–2139.

Newman, E.M. & Santi, D.V. (1982) Proc. Natl. Acad. Sci. USA *79*, 6419–6423.

Newport, J. & Kirschner, M. (1982) Cell *30*, 687–696.

Nevers, P. & Spatz, H.C. (1975) Mol. Gen. Genet. *139*, 233–248.

Nickol, J.M. & Felsenfeld, G. (1983) Cell *35*, 467–477.

Nickol, J., Behe, M. & Felsenfeld, G. (1982) Proc. Natl. Acad. Sci. USA *79*, 1771–1775.

Niwa, O. & Sugahara, T. (1981) Proc. Natl. Acad. Sci. USA *78*, 6290–6294.

Niwa, O., Yokota, Y., Ishida, H. & Sugahara, T. (1983) Cell *32*, 1105–1113.

Noda, M., Teranishi, Y., Takahashi, H., Toyosato, M., Notake, M., Nakanishi, S. & Numa, S. (1982) Nature, *297*, 431–434.

Noguchi, H., Reddy, G.P. & Pardee, A.B. (1983) Cell *32*, 443–451.

Nordheim, A. & Rich, A. (1983) Proc. Natl. Acad. Sci. USA *80*, 1821–1825.

Nordheim, A., Pardue, M.L., Lafer, E.M., Moller, A., Stollar, B.D. & Rich, A. (1981) Nature *294*, 417–422.

Nordheim, A., Lafer, E.M., Peck, L.J., Wang, J.C., Stollar, B.D. & Rich, A. (1982) Cell *31*, 309–318.

O'Connor, C.D. & Humphreys, G.O. (1982) Gene *20*, 219–229.

Oden, K.L. & Clarke, S. (1983) Biochemistry *22*, 2978–2986.

Okada, G., Teraoka, H. & Tsukada, K. (1981) Biochemistry *20*, 934–940.

Olsson, L. & Forchhammer, J. (1984) Proc. Natl. Acad. Sci. USA *81*, 3389–3393.

Orkin, S.H. (1983) Nature *301*, 108–109.

Ormondt, van, H., Garter, J., Havelaar, K.J. & de Waard, A. (1975) Nucl. Acids Res. *2*, 1391–1400.

Ostrander, M., Vogel, S. & Silverstein, S. (1982) Mol. Cell Biol. *2*, 708–714.

Ott, M.O., Sperling, L., Cassio, D., Levilliers, J., Sala-Trepat, J. &,Weiss, M. (1982) Cell *30*, 825–833.

Pages, M. & Roizes, G. (1982) Nucl. Acids Res. *10*, 565–576.

Paik, W.K. & Kim, S. (1975) Adv. Enzymol. *42*, 227–286.

Pakhomova, M.V., Zaitseva, G.N. & Belozerskii, A.N. (1968) Dokl. Acad. Nauk. SSSR *182*, 712–714.

Palmiter, R.D., Chen, H.Y., & Brinster, R.L. (1982) Cell *29*, 701–710.

Palmiter, R.D., Wilkie, T.M., Chen, H Y. & Brinster, R.L. (1984) Cell *36*, 869–877.

Papioannou, V.E., McBurney, M.W., Gardiner, R.L. & Evans, M.J. (1975) Nature *258*, 70–73.

Parker, M.I., Judge, K. & Gevers, W. (1982) Nucl. Acids Res. *10*, 5879–5891.

Parks, L.W. & Schlenk, F. (1958) J. Biol. Chem. *230*, 295–305.

Pays, E., Delauw, M.F., Laurent, M. & Steinert, M. (1984) Nucl. Acids Res. *12*. 5235–5247.

Pech, M., Streeck, R.E. & Zachau, H.G. (1979) Cell *15*, 883–893.

Peck, L.J. & Wang, J.C. (1983) Proc. Natl. Acad. Sci. USA *80*, 6206–6210.

Pedersen, R.A. & Spindle, A.I. (1980) Nature *284*, 550–552.

Pegg, A.E. (1971) FEBS Letts. *16*, 13–16.

———. (1974) Biochem. J. *141*, 581–583.

Pellicer, A., Robins, D., Wold, B., Sweet, R., Jackson, J., Lowry, I., Roberts, J.M., Sim, G.K., Silverton, S. & Axel, R. (1980) Science *209*, 1414–1422.

Pennock, D.G. & Reeder, R.H. (1984) Nucl. Acids Res. *12*, 2225–2232.

Peoples, D.P. & Hardman, N. (1983) Nucl. Acids Res. *11*, 7777–7788.

Perry, R.P. (1984) Nature *310*, 14–15.

Perry, R.P. & Kelley, D.E. (1974) Cell *1*, 37–42.

Pettijohn, D.E. & Pfenninger, O. (1980) Proc. Natl. Acad. Sci. USA *77*, 1331–1335.

Pfeifer, G.P., Grunwald, S., Boehm, T.L.J. & Drahovsky, D. (1983) Biochim. Biophys. Acta *740*, 323–330.

Pfohl-Leszkowicz, A., Salas, C., Fuchs, R.P.P. & Dirheimer, G. (1981) Biochemistry *20*, 3020–3024.

Pfohl-Leszkowicz, A., Galiegue-Zouitina, S., Bailleul, B., Loucheux-Lefebvre, M.H. & Dirheimer, G. (1983) FEBS Letts. *163*, 85–88.

Plasterk, R.H.A., Vrieling, H. & Van der Putte, P. (1983) Nature *301*, 344–347.

Plasterk, R.H.A., Vollering, M., Brinkman, A. & Van der Putte, P. (1984) Cell *36*, 189–196.

Pohl, F.M. & Jovin, T.M. (1972) J. Mol. Biol. *67*, 375–396.

Pohl, F.M., Thomas, R. & Di Capua, E. (1982) Nature *300*, 545–546.

Pollack, Y., Stein, R., Razin, A. & Cedar, H. (1980) Proc. Natl. Acad. Sci. USA *77*, 6463–6467.

Pollack, Y., Kasir, J., Shemer, T. Metzger, S. & Szyf, M. (1984) Nucl. Acids Res. *12*, 4811–4824.

Pollock, J.M., Swihart, M. & Taylor, J.H. (1978) Nucl. Acids Res. *5*, 4855–4863.

Ponzetto-Zimmerman, C. & Wolgemuth, D.J. (1984) Nucl. Acids Res. *12*, 2807–2822.

Poschl, E. & Streek, R.E. (1980) J. Mol. Biol. *143*, 147–154.

Pratt, K. & Hattman, S. (1981) Mol. Cell Biol. *1*, 600–608.

Proffitt, J.H., Davie, J.R., Swinton, D. & Hattman, S. (1984) Mol. Cell Biol. *4*, 985–988.

Pukkila, P.J., Peterson, J., Herman, G., Modrich, P. & Meselson, M. (1983) Genetics *104*, 571–582.

Quadrifoglio, F., Manzini, G., Vasser, M., Dinkelspiel, K. & Crea, R. (1981) Nucl. Acids Res. *9*, 2195–2206.

Quint, A. & Cedar, H. (1981) Nucl. Acids Res. *9*, 633–646.

Rabbitts, T.H. (1983) Mol. Biol. Med. *1*, 275–281.

Radman, M. (1981) Biochem. Soc. Trans. *9*, 78p.

Radman, M. Villani, G., Boiteux, S., Kinsella, A.R., Glickman, B.W. & Spadari, S. (1978) Cold Spring Harbor Symp. Quant. Biol. *43*, 937–945.

Radman, M., Dohet, C., Bourguiguon, M-F., Doubleday, O.P. & Lecomte, P. (1981) In Chromosome Damage and Repair, 431–445, Seeberg, E. & Kleppe, K. eds. Plenum, New York.

Rae, P.M.M. (1976) Science *194*, 1062–1064.

Rae, P.M.M. & Spear, B.B. (1978) Proc. Natl. Acad. Sci. USA *75*, 4992–4996.

Rae, P.M.M. & Steele, R.E. (1979) Nucl. Acids Res. *6*, 2987–2995.

Raibaud, A., Gaillard, C., Longacre, S., Hibner, U., Buck, G., Bernardi, G. & Eisen, H. (1983) Proc. Natl. Acad. Sci. USA *80*, 4306–4310.

Ray, D.S. & Hanawalt, P.C. (1964) J. Mol. Biol. *9*, 812–824.

Razin, A. (1973) Proc. Natl. Acad. Sci. USA *70*, 3773–3775.

Razin, A. & Cedar, H. (1977) Proc. Natl. Acad. Sci. USA *74*, 2725–2728.

Razin, A. & Friedman, J. (1981) Prog. in Nucl. Acids Res. and Mol. Biol. *25*, 33–52.

Razin, A. & Razin, S. (1980) Nucl. Acids Res. *8*, 1383–1390.

Razin, A. & Riggs, A.D. (1980) Science *210*, 604–610.

Razin, A., Sedat, J.W. & Sinsheimer, R.L. (1973) J. Mol. Biol. *78*, 417–425.

Razin, A., Uriel, S., Pollock, Y., Gruenbaum, Y. & Glaser, G. (1980) Nucl. Acids Res. *8*, 1783–1792.

Razin, A., Webb, C., Szyf, M., Yisraeli, J., Rosenthal, A., Naveh-Many, T., Sciaky-Gallili, N. & Cedar, H. (1984) Proc. Natl. Acad. Sci. USA *81*, 2275–2279.

Razin, A., Cedar, H. & Riggs, A.D. (1984) *DNA Methylation; Biochemistry and Biological Significance*. Springer-Verlag, New York.

Razin, S.V., Mantieva, V.L. & Georgiev, G.P. (1978) Nucl. Acids Res. *5*, 4737–4751.

Reichman, M. & Penman, S. (1973) Biochim. Biophys. Acta *324*, 282–289.

Reilly, J.G., Thomas, C.A. & Sen, A. (1982) Biochim. Biophys. Acta *697*, 53–59.

Reis, R.J.S. & Goldstein, S. (1982a) Proc. Natl. Acad. Sci. USA *79*, 3949–3953.

———. (1982b) Nucl. Acids Res. *10*, 4293–4304.

———. (1983) J. Biol. Chem. *258*, 9078–9085.

Reiser, J. & Yuan, R. (1977) J. Biol. Chem. *252*, 451–456.

Reitz, M.S., Mann, D.L., Eiden, M., Trainor, C.D. & Clarke, M.F. (1984) Mol. Cell Biol. *4*, 890–897.

Rhodes, D. & Klug, A. (1981) Nature *29*, 378–380.

Rich, A. (1982) Cold Spring Harbor Symp. Quant. Biol. 47, 1–12.

Riggs, A.D. (1975) Cytogenet. Cell Genet. *14*, 9–25.

Roberts, R.J. (1980) Gene *8*, 329–343.

———. (1982) Nucl. Acids Res. *10*, r117–r144.

———. (1984) Nucl. Acids Res. *12* (Suppl.), r167–r204.

Rogers, J. & Wall, R. (1981) Proc. Natl. Acad. Sci. USA *78*, 7497–7501.

Roginski, R.S., Scoultchi, A.I., Henthorn, P., Smithies, O., Hsiung, N. & Kucherlapati, R. (1983) Cell *35*, 149–155.

Romanov, G.A. & Vanyushin, B.F. (1980) Mol. Biol. *3*, 279–288.

———. (1981) Biochim. Biophys. Acta *653*, 204–218.

Rossant, J. (1977) In Development in Mammals, vol. 2: 119–150, Johnson, M.H., ed. Elsevier/North Holland, Amsterdam.

Roy, P.H. & Smith, H.O. (1973) J. Mol. Biol. *81*, 445–459.

Roy, P.H. & Weissbach, A. (1975) Nucl. Acids Res. *2*, 1669–1684.

Royer, H.D. & Sager, R. (1979) Proc. Natl. Acad. Sci. USA *76*, 5794–5798.

Rubery, E.D. & Newton, A.A. (1973) Biochim. Biophys. Acta *324*, 24–36.

Rubin, R.A. & Modrich, P. (1977) J. Biol. Chem. *252*, 7265–7272.

Ruchirawat, M., Becker, F.F. & Lepeyre, J-N. (1984) Nucl. Acids Res. *12*, 3357–3372.

Russell, G.J., Walker, P.M.B., Elton, R.A. & Subak-Sharpe, J.H. (1976) J. Mol. Biol. *108*, 1–23.

Rydberg, B. & Lindahl, T. (1982) EMBO J. *1*, 211–216.

Rydberg, S. (1977) Mol. Gen. Genet. *152*, 19–28.

Ryoji, M. & Worcel, A. (1984) Cell *37*, 21–32.

Sager, R. (1954) Proc. Natl. Acad. Sci. USA *40*, 356–363.

Sager, R. & Grabowy, C. (1983) Proc. Natl. Acad. Sci. USA *80*, 3025–3029.

Sager, R. & Kitchin, R. (1975) Science *189*, 426–433.

Sager, R. & Lane, D. (1972) Proc. Natl. Acad. Sci. USA *69*, 2410–2413.

Sager, R. & Ramanis, Z. (1973) Theor. Appl. Genet. *43*, 101–108.

———. (1974) Proc. Natl. Acad. Sci. USA *71*, 4698–4702.

Sager, R., Sano, H. & Grabowy, C.T. (1984) In Curr. Top. Microbiol. and Immunol., vol. 108: 157–172, Methylation of DNA, Trautner, T.A., ed. Springer-Verlag, Berlin.

Sain, B. & Murray, N.E. (1980) Mol. Gen. Genet. *180*, 35–46.

Sakano, H., Huppi, K., Heinrich, G. & Tonegawa, S. (1979) Nature *280*, 288–294.

Salas, C.E., Pfohl-Leszkowicz, A., Lang, M.C. & Dirheimer, G. (1979) Nature *278*, 71–72.

Salser, W. (1977) Cold Spring Harbor Symp. Quant. Biol. *42*, 985–1002.

Sanders, M.M. (1978) J. Cell Biol. *79*, 97–109.

Sanford, J., Forrester, L., Chapman, V., Chandley, A. & Hastie, N. (1984) Nucl. Acids Res. *12*, 2823–2836.

Sano, H. & Sager, R. (1980) Eur. J. Biochem. *105*, 471–480.

———. (1982) Proc. Natl. Acad. Sci. USA *79*, 3584–3585.

Sano, H., Royer, H-D. & Sager, R. (1980) Proc. Natl. Acad. Sci. USA *77*, 3581–3585.

Sano, H., Grabowy, C. & Sager, R. (1981) Proc. Natl. Acad. Sci. USA *78*, 3118–3122.

Sano, H., Naguchi, H. & Sager, R. (1983) Eur. J. Biochem. *135*, 181–185.

Santi, D.V., Garrett, C.E. & Barr, P.J. (1983) Cell *33*, 9–10.

Santi, D.V., Norment, A. & Garrett, C.E. (1984) Proc. Natl. Acad. Sci. USA *81*, 6993–6997.

Scarano, E. (1969) Ann. Embryol. Morphogen *1*, (Suppl.) 51–61.

Scarano, E., Iaccarino, M., Grippo, P. & Parisi, E. (1967) Proc. Natl. Acad. Sci. USA *57*, 1394–1400.

Scarbrough, K., Hattman, S. & Nur, U. (1984) Mol. Cell Biol. *4*, 599–603.

Schildkraut, C.L., Marmur, J. & Doty, P. (1962) J. Mol. Biol. *4*, 430–443.

Schlagman, S.L. & Hattman, S. (1983) Gene *22*, 139–156.

Schlenk, F. (1977) In The Biochemistry of Adenosylmethionine, 3–17, Salvatore, F., Borek, E., Sappia, V., Williams-Ashman, H.G. & Schlenk, F., eds. Columbia University Press, New York.

Schlenk, F., Dainko, J.L. & Stanford, S.M. (1959) Arch. Biochem. Biophys. *83*, 28–34.

Schlenk, F., Zydek, C.R., Ehninger, D.J. & Dainko, J.L. (1965) Enzymologia *29*, 283–298.

Schneiderman, M.H. & Billen, D. (1973) Biochim. Biophys. Acta *308*, 352–360.

Searle, P.F. & Tata, J.R. (1981) Cell *23*, 741–746.

Searle, S., Gillespie, D.A.F., Chiswell, D.J. & Wyke, J.A. (1984) Nucl. Acids Res. *12*, 5193–5210.

Sekikawa, K. & Levine, A.J. (1981) Proc. Natl. Acad. Sci. USA *78*, 1100–1104.

Setlow, P. (1976) In Handbook of Biochemistry and Molecular Biology: Nucleic Acids, vol. 2: 312, Fasman, G.D., ed. CRC Press, Cleveland, OH.

Shapiro, H.S. (1976) In Handbook of Biochemistry and Molecular Biology: Nucleic Acids, vol. 2: 259, Fasman, G.D., ed. CRC Press, Cleveland, OH.

Shapiro, L.J. & Mohandas, T. (1983) Cold Spring Harbor Symp. Quant. Biol. *47*, 631–637.

Shapiro, R., Brauerman, B., Louis, J.B. & Servis, R.E. (1973) J. Biol. Chem. *248*, 4060–4064.

Sheffery, M., Rifkind, R.A. & Marks, P.A. (1982) Proc. Natl. Acad. Sci. USA *79*, 1180–1184.

Shein, A., Berdahl, B.J., Low, M. & Borek, E. (1972) Biochim. Biophys. Acta *272*, 481–485.

Shen, C-K. J. & Maniatis, T. (1980) Proc. Natl. Acad Sci. USA *77*, 6634–6638.

Shen, S., Slightom, J.L. & Smithies, O. (1981) Cell *26*, 191–203.

Shirayoshi, Y., Okada, T.S. & Takeichi, M. (1983) Cell *35*, 631–638.

Simon, D., Grunert, F., v. Acken, U., Doring, H.P. & Kroger, H. (1978) Nucl. Acids Res. *5*, 2153–2167.

Simon, D., Grunert, F., Kroger, H. & Grassman, A. (1980) Eur. J. Cell Biol. *22*, 33.

Simon, D., Stuhlmann, H., Jahner, D., Wagner, H., Werner, E. & Jaenisch, R. (1983) Nature *304*, 275–277.

Sinden, R.R., Carlson, J.P. & Pettijohn, D.E. (1980) Cell *21*, 773–783.

Singer, J., Stellwagen, R.H., Roberts-Ems, J. & Riggs, A.D. (1977) J. Biol. Chem. *252*, 5509–5513.

Singer, J., Roberts-Ems, J. & Riggs, A.D. (1979) Science *203*, 1019–1021.

Singer, J., Roberts-Ems, J., Luthardt, F.W. & Riggs, A.D. (1979) Nucl. Acids Res. *7*, 2369–2385.

Singer, J., Schmute, W.C., Shively, J.E., Todd, C.W., & Riggs, A.D. (1979) Anal. Biochem. *94*, 297–301.

Singleton, C.K., Klysik, J., Stirdivant, S.M. & Wells, R.D. (1982) Nature *299*, 312–316.

Singleton, C.K., Klysik, J. & Wells, R.D. (1983) Proc. Natl. Acad. Sci. USA *80*, 2447–2451.

Sinsheimer, R.L. (1955) J. Biol. Chem. *215*, 579–583.

Slack, J.M.W. (1980) Nature *286*, 760.

Slightom, J.L., Blechl, A.E. & Smithies, O. (1980) Cell *21*, 627–638.

Small, D., Nelkin, B. & Vogelstein, B. (1982) Proc Natl. Acad. Sci. USA *79*, 5911–5915.

Smith, D.W., Garland, A.M., Herman, G., Enns, R.E., Baker, T.A. & Zyskind, J.W. (1985) EMBO Journal (in press).

Smith, J.D. (1976) Biochem. Biophys. Res. Comm. *73*, 7–12.

Smith, S.S., Yu, J.C. & Chen, C.W. (1982) Nucl. Acids Res. *10*, 4305–4320.

Smith, T.F., Waterman, M.S. & Sadler, J.R. (1983) Nucl. Acids Res. *11*, 2205–2220.

Sneider, T.W. (1972) J. Biol. Chem. *247*, 2872–2875.

———. J. Mol. Biol. *79*, 731–734.

———. (1980) Nucl. Acids Res. *8*, 3829–3840.

Sneider, T.W., Teague, W.M. & Rogachevsky, L.M. (1975) Nucl. Acids Res. *2*, 1685–1700.

Solage, A. & Cedar, H. (1978) Biochemistry *17*, 2934–2938.

Spadafora, C. & Crippa, M. (1984) Nucl. Acids Res. *12*, 2691–2704.

Spoerel, N., Herrlich, P. & Bickle, T.A. (1979) Nature *278*, 30–34.

Stafford, J. & Queen, C. (1983) Nature *306*, 77–79.

Stein, R., Gruenbaum, Y., Pollack, Y., Razin, A. & Cedar, H. (1982a) Proc. Natl. Acad. Sci. USA *79*, 61–65.

Stein, R., Razin, A. & Cedar, H. (1982b) Proc. Natl. Acad. Sci. USA *79*, 3418–3422.

Stein, R., Sciaky-Gallili, N., Razin, A. & Cedar, H. (1983) Proc. Natl. Acad. Sci. USA *80*, 2422–2426.

Stewart, C.L., Stuhlman, H., Jahner, D., & Jaenisch, R. (1982) Proc. Natl. Acad. Sci. USA *79*, 4098–4102.

Stolowicz, M.L. & Minch, J.J. (1981) J. Am. Chem. Soc. *103*, 6015–6019.

Storb, U. & Arp, B. (1983) Proc. Natl. Acad. Sci. USA *80*, 6642–6646.

Struhl, G. (1984) Nature *310*, 10–11.

Stuhlmann, H., Jahner, D. & Jaenisch, R. (1981) Cell *26*, 221–232.

Sturm, K.S. & Taylor, J.H. (1981) Nucl. Acids Res. *9*, 4537–4546.

Subramanian, K.N. (1982) Nucl. Acids Res. *10*, 3475–3486.

Sugimoto, K., Oka, A., Sugisaki, H., Takanami, M., Nishimura, A., Yasuda, Y. & Hirota, Y. (1979) Proc. Natl. Acad. Sci. USA *76*, 575–579.

Suri, B., Nagaraja, V. & Bickle, T.A. (1984) Curr. Top. Microbiol. and Immunol. *108*, 1–9.

Sutter, D. & Doerfler, W. (1980) Proc. Natl. Acad. Sci. USA *77*, 253–256.

Svihla, E. & Schlenk, F. (1960) J. Bacteriol. *79*, 841–848.

Swinton, D., Hattman, S., Crain, P.F., Cheng, C-S., Smith, D.L. & McCloskey, J.A. (1983) Proc. Natl. Acad. Sci. USA *80*, 7400–7404.

Szer, W. & Shugar, D. (1966) J. Mol. Biol. *17*, 174–187.

Szomolanyi, E., Kiss, A. & Venetianer, P. (1980) Gene *10*, 219–225.

Szyf, M., Gruenbaum, Y., Urieli-Shoval, S. & Razin, A. (1982) Nucl. Acids Res. *10*, 7247–7259.

Szyf, M., Avraham-Haetzni, K., Reifman, A., Shlomai, J., Kaplan, F., Oppenheim, A. & Razin, A. (1984) Proc. Natl. Acad. Sci. USA *81*, 3278–3282.

Tabor, H. & Tabor, C.W. (1972) Adv. Enzymol. Relat. Areas Mol. Biol. *36*, 203–268.

Takagi, N., Yoshida, M.A., Sugawara, O. & Sasaki, M. (1983) Cell *34*, 1053–1062.

Takahashi, I. & Marmur, J. (1963) Nature *197*, 794–795.

Talwar, S., Pocklington, M.J. & Maclean, N. (1984) Nucl. Acids Res. *12*, 2509–2517.

Tamame, M., Antequera, F., Villanueva, J.R. & Santos, T. (1983) Mol. Cell Biol. *3*, 2287–2297.

Tanaka, M., Hibasami, H., Nagai, J. & Ikeda, T. (1980) Aus. J. Exp. Biol. Med. Sci. *58*, 391–396.

Tanaka, K., Appella, E. & Jay, G. (1983) Cell *35*, 457–465.

Tanford, C. (1961) Physical Chemistry of Macromolecules, Wiley, New York.

Tantravahi, U., Guntaka, R.V., Erlanger, B.F. & Miller, O.J. (1981) Proc. Natl. Acad. Sci. USA *78*, 489–493.

Taparowsky, E.J. & Gerbi, S.A. (1982) Nucl. Acids Res. *10*, 1271–1281.

Taylor, J.H. (1978) In DNA Synthesis: Present and Future, 143–159, Molineux, I. & Kohiyama, M., eds. Plenum, New York.

Taylor, J.H. (1979) Molecular Genetics III p 89–115. Academic Press, New York.

Taylor, J.H. (1984) *DNA Methylation and Cellular Differentiation*, Springer-Verlag, New York.

Taylor, S.M. & Jones, P.A. (1979) Cell *17*, 771–779.

————. (1982) J. Mol. Biol. *162*, 679–692.

Taylor, S.M., Constantinides, P.A. & Jones, P.A. (1984) Curr. Top. Microbiol. and Immunol. *108*, 115–127.

Theiss, G. & Follmann, H. (1980) Biochem. Biophys. Res. Comm. *94*, 291–297.

Thiery, J-P., Ehrlich, S.D., Devillers-Thiery, A. & Bernardi, G. (1973) Eur. J. Biochem. *38*, 434–442.

Thomae, R., Beck, S. & Pohl, F.M. (1983) Proc. Natl. Acad. Sci. USA *80*, 5550–5553.

Thomas, A.J. & Sherratt, H.S.A. (1956) Biochem. J. *62*, 1–4.

Toniolo, D., D'urso, M., Mastini, G., Persico, M., Tufano, V., Battistuzzi, G. & Luzzatto, L. (1984) EMBO Journal *3*, 1987–1995.

Tosi, L., Granieri, A. & Scarano, E. (1972) Exptl. Cell Res. *72*, 257–264.

Toussaint, A. (1976) Virology *70*, 17–27.

Trautner, T.A., ed. (1984) Current Topics in Microbiology and Immunology, vol. 108.

Trifonov, E.N. (1982) Cold Spring Harbor Symp. Quant. Biol. 47, 271–278.

Turkington, R.W. & Spielvogel, R.L. (1971) J. Biol. Chem. *246*, 3835–3840.

Turnbull, J.F. & Adams, R.L.P. (1976) Nucl. Acids Res. *3*, 677–695.

Tykocinski, M.L. & Max, E.E. (1984) Nucl. Acids Res. *12*, 4385–4396.

Ueland, P.M. (1982) Pharmacol. Rev. *34*, 223–253.

Ueland, P.M., Helland, S., Brach, O.J., Schanche, J-S. (1984) J. Biol. Chem. *259*, 2360–2364.

Ullrich, A., Dull, T.J., Gray, A., Philips, J.A. & Peter, S. (1982) Nucl. Acids Res. *10*, 2225–2240.

Urieli-Shoval, S., Gruenbaum, Y., Sedat, J. & Razin, A. (1982) FEBS Letts. *146*, 148–152.

van der Ploeg, L.H.T. & Flavell, R.A. (1980) Cell *19*, 947–958.

van der Ploeg, L.H.T., Groffen, J. & Flavell, R.A. (1980) Nucl. Acids Res. *8*, 4563–4574.

van der Vliet, P.C. & Levine, A.J. (1973) Nature (New Biol.) *246*, 170–174.

van Ormondt, H. *See* Ormondt, van, H.

Vanyushin, B.F. (1984) Curr. Top. Microbiol. and Immunol. *108*, 99–114.

Vanyushin, B.F. & Belozersky, A.N. (1959) Dokl. Acad. Nauk. SSSR *193*, 1422–1425.

Vanyushin, B.F. & Kirnos, M.D. (1974) FEBS Letts. *39*, 195–199.

————. (1977) Biochim. Biophys. Acta *475*, 323–336.

Vanyushin, B.F. & Milner, B. (1965) Nauchn. Dokl. Vissh. Schk. *2*, 162.

Vanyushin, B.F., Belozersky, A.N., Kokurina, N.A. & Kodirovo, D.X. (1968) Nature *218*, 1066–1067.

Vanyushin, B.F., Tkacheva, B.G. & Belozersky, A.N. (1970) Nature *225*, 948–949.

Vanyushin, B.F., Nemirovsky, L.E., Klimenki, V.V., Vasiliev, V.K. & Belozersky, A.N. (1973) Gerontologia *19*, 138–152.

Vanyushin, B.F., Mazin, A.L., Vasiliev, V.K. & Belozersky, A.N. (1973) Biochim. Biophys. Acta *299*, 397–403.

Vardimon, L. & Rich, A. (1984) Proc. Natl. Acad. Sci. USA *81*, 3268–3272.

Vardimon, L., Neumann, T. Kuhlmann, I., Sutter, D. & Doerfler, W. (1980) Nucl. Acids Res. *8*, 2461–2469.

Vardimon, L., Gunthert, U. & Doerfler, W. (1982) Mol. Cell Biol. *2*, 1574–1580.

Vardimon, L., Kressmann, A., Cedar, H., Maechler, M. & Doerfler, W. (1982) Proc. Natl. Acad. Sci. USA *79*, 1073–1077.

Vedel, M., Gomez-Garcia, M., Sala, M. & Sala-Trepat, J.M. (1983) Nucl. Acids Res. *11*, 4335–4354.

Venolia, L. & Gartler, S.M. (1983a) Nature *302*, 82–83.

———. (1983b) Somatic Cell Genet. *9*, 617–627.

Vesely, J. & Cihak, A. (1973) Experientia *29*, 1132–1133.

———. (1977) Cancer Res. *37*, 3684–3689.

———. (1978) Pharmacol. Ther. A. *2*, 813–840.

Vesely, J., Cihak, A. & Sorm, F. (1969) Collect. Czech. Chem. Comm. *34*, 901–909.

Viegas-Pequignot, E., Derbin, C., Malfoy, B., Taillandier, A., Leng, M. & Dutrillaux, B. (1983) Proc. Natl. Acad. Sci. USA *80*, 5890–5894.

Vincent, B.R. de S. & Wahl, G.M. (1983) Proc. Natl. Acad. Sci. USA *80*, 2002–2006.

Vovis, G.F., Horiuchi, K. & Zinder, N.D. (1974) Proc. Natl. Acad. Sci. USA *71*, 3810–3813.

Waechter, D.E. & Baserga, R. (1982) Proc. Natl. Acad. Sci. USA *79*, 1106–1110.

Wagner, I. & Capesius, I. (1981) Biochim. Biophys. Acta *654*, 52–56.

Wagner, R.E. & Meselson, M. (1976) Proc. Natl. Acad. Sci. USA *73*, 4135–4139.

Walder, R.Y., Hartley, J.L. Donelson, J.E. & Walder, J.A. (1981) Proc. Natl. Acad. Sci. USA *78*, 1503–1507.

Walder, R.Y., Langtimm, C.J., Chatterjee, R. & Walder, J.A. (1983) J. Biol. Chem. *258*, 1235–1241.

Wang, J.C. (1979) Proc. Natl. Acad. Sci. USA *76*, 200–203.

Wang, R.Y-H., Gehrke, C.W. & Ehrlich, M. (1980) Nucl. Acids Res. *8*, 4777–4790.

Wang, R.Y-H., Kuo, K.C., Gehrke, C.W., Huang, L-H. & Ehrlich, M. (1982) Biochim. Biophys. Acta *697*, 371–377.

Wang, R.Y-H., Huang, L-H. & Ehrlich, M. (1984) Nucl. Acids Res. *12*, 3473–3489.

Warner, A.H. & Bagshaw, J.C. (1984) Dev. Biol. *102*, 264–267.

Watabe, H., Iino, T., Kaneko, T., Shibata, T. & Ando, T. (1983) J. Biol. Chem. *258*, 4663–4665.

Watabe, H., Shibata, T., Iino, T. & Ando, T. (1984) J. Biochem. *95*, 1677–1690.

Weatherall, D.J. & Clegg, J.B. (1979) Cell *16*, 467–479.

Weintraub, H. (1979) Nucl. Acids Res. *7*, 761–792.

Weintraub, H., Larsen, A. & Groudine, M. (1981) Cell *24*, 333–344.

Weisbrod, S.T. (1982) Nucl. Acids Res. *10*, 2017–2042.

Weiss, J.W. & Pitot, H.C., (1974) Arch. Biochem. Biophys. *165*, 588–596.

White, R. & Parker, M. (1983) J. Biol. Chem. *258*, 8943–8948.

Whitfield, P.R. & Spencer, D. (1968) Biochim. Biophys. Acta *157*, 333–343.

Whittaker, P.A., McLachlan, A. & Hardman, N. (1981) Nucl. Acids Res. *9*, 801–814.

Wickner, R.B., Tabor, C.W. & Tabor, H. (1970) J. Biol. Chem. *245*, 2132–2139.

Wigler, M., Sweet, R., Sim, G.K., Wold, B., Pellicer, A., Lacy, E., Maniatis, T., Silverstein, S. & Axel, R. (1979) Cell *16*, 777–785.

Wigler, M., Levy, D. & Perucho, M. (1981) Cell *24*, 33–40.

Wilkie, N.M., Clements, J.B., Boll, W., Mantei, N., Lonsdale, D. & Weissman, C. (1980) Nucl. Acids Res. *8*, 859–877.

Wilks, A.F., Cozens, P.J., Mattaj, I.W. & Jost, J-P. (1982) Proc. Natl. Acad. Sci. USA *79*, 4252–4255.

Wilks, A., Seldran, M. & Jost, J-P. (1984) Nucl. Acids Res. *12*, 1163–1177.

Williams-Ashman, H.G. & Pegg, A.E. (1980) In Polyamines in Biology and Medicine, Morris, D.R. & Marton, L., eds. Marcel Dekker, New York.

Williams-Ashman, H.G., Corti, A. & Coppoc, G.L. (1977) In The Biochemistry of Adenosylmethionine, 473–489, Salvatore, F., Borek, E., Zappia, V., Williams-Ashman, H.G. & Schlenk, F., eds. Columbia University Press, New York.

Willis, D.B. & Granoff, A. (1980) Virology *107*, 250–257.

Willis, D.B., Goorha, R. & Granoff, A. (1984) J. Virol. *49*, 86–91.

Wilson, V.L. & Jones, P.A. (1983) Cell *32*, 239–246.

Wilson, V.L., Jones, P.A. & Momparler, R.L. (1983) Cancer Res. *43*, 3493–3496.

Wilson, M.J., Shivapurkar, N. & Poirier, L.A. (1984) Biochem. J. *218*, 987–990.

Wolf, S.F. & Migeon, B.R. (1982) Nature *295*, 667–671.

———. (1983) Cold Spring Harbor Symp. Quant. Biol. *47*, 621–630.

———. (1985) Nature *314*, 467–469.

Wolf, S.F., Jolly, D.J., Lunnen, K.D., Friedmann, T. & Migeon, B.R. (1984a) Proc. Natl. Acad. Sci. USA *81*, 2806–2810.

Wolf, S.F., Dintzis, S., Toniolo, D., Persico, G., Lunnen, K.D., Axelman, J. & Migeon, B.R. (1984b) Nucl. Acids Res. *12*, 9333–9348.

Wood, W.G., Bunch, C., Kelly, S., Gunn, Y. & Breckon, G. (1985) Nature *313*, 320–323.

Woodbury, C.P., Downey, R.L. & von Hippel, P.H. (1980) J. Biol. Chem. *255*, 11526–11533.

Woodcock, D.M., Adams, J.K. & Cooper, I.A. (1982) Biochim. Biophys. Acta *696*, 15–22.

Woodcock, D.M., Adams, J.K., Allan, R.G. & Cooper, I.A. (1983) Nucl. Acids Res. *11*, 489–499.

Woodcock, D.M., Crowther, P.J. Simmons, D.L. & Cooper, I.A. (1984) Biochim. Biophys. Acta *783*, 227–233.

Wright, S., Rosenthal, A., Flavell, R. & Grosveld, F. (1984) Cell *38*, 265–273.

Wu, S-E., Huskey, W.P., Borchardt, R.T. & Schowen, R.L. (1983) Biochemistry *22*, 2828–2832.

Wyatt, G.R. (1951) Biochem. J. *48*, 584–590.

Wyatt, G.R. & Cohen, S.S. (1953) Biochem. J. *55*, 774–782.

Yen, P.H., Patel, P., Chinault, A.C., Mohandas, T. & Shapiro, L.J. (1984) Proc. Natl. Acad. Sci. USA *81*, 1759–1763.

Yoo, O.J. & Agarwal, L.K. (1980) J. Biol. Chem. *255*, 6445–6449.

Yoo, O.J., Dwyer-Hallquist, P. & Agarwal, K.L. (1982) Nucl. Acids Res. *10*, 6511–6519.

Young, E.T. & Sinsheimer, R.L. (1965) J. Biol. Chem. *240*, 1274–1280.

Young, P.R. & Tilghman, S.M. (1984) Mol. Cell Biol. *4*, 898–907.

Yuan, R. (1981) Ann. Rev. Biochem. *50*, 285–315.

Yuan, R., Bickle, T.A., Ebbers, W. & Brack, C. (1975) Nature *256*, 556–560.

Yuan, R., Hamilton, D.L. & Burckhardt, J. (1980) Cell *20*, 237–244.

Zadrazil, S., Fucik, V., Bartl, P., Sarmova, Z. & Sorm, F. (1965) Biochim. Biophys. Acta *108*, 701–703.

Zain, B.S., Adams, R.L.P. & Imrie, R.C. (1973) Cancer Res. *33*, 40–46.

Zappia, V., Zydek-Cwick, C.R. & Schlenk, F. (1969) J. Biol. Chem. *244*, 4499–4509.

Zappia, V., Carteni-Farina, M. & Galletti, P. (1977a) In The Biochemistry of Adenosylmethionine, 493–509, Salvatore, F., Borek, E., Zappia, V., Williams-Ashman, H.G. & Schlenk, F., eds. Columbia University Press, New York.

Zappia, V., Galletti, P., Olivia, A. & DeSantis, A. (1977b) Anal. Biochem. *79*, 535–543.

Zappia, V., Carteni-Farina, M., & Porcelli, M. (1979) in Transmethylation, 95–104, Usdin, E., Borchardt, R.T. & Creveling, C.R., eds. Elsevier/North Holland, New York.

Zimmerman, S.B. (1982) Ann. Rev. Biochem. *51*, 359–428.

Zyskind, J.W. & Smith, D.W. (1980) Proc. Natl. Acad. Sci. USA *77*, 2460–2464.

Index